D1141336

NEW HOLLAND PROFESSIONAL

PLUMBING

NEW HOLLAND PROFESSIONAL

PLUMBING

TONY BEAUMONT

First published in 2007 by New Holland Publishers (UK) Ltd
London • Cape Town • Sydney • Auckland

Garfield House
86–88 Edgware Road
London W2 2EA
United Kingdom
www.newhollandpublishers.com

80 McKenzie Street
Cape Town 8001
South Africa

Unit 1
66 Gibbes Street
Chatswood
NSW 2067
Australia

218 Lake Road
Northcote
Auckland
New Zealand

ISBN 978 1 84537 721 2

Senior Editor: Corinne Masciocchi
Design: www.bluegumdesigners.com
Production: Hazel Kirkman
Editorial Direction: Rosemary Wilkinson

2 4 6 8 10 9 7 5 3 1

Reproduction by Pica Digital Pte Ltd, Singapore
Printed and bound by Craft Print International Ltd, Singapore

Contents

Introduction 6

TOOLS 8

TECHNIQUES 18

Earthing 20

Safety and preparation 22

Pipe materials 23

Fittings 25

Cutting pipework 28

Joining methods 31

Isolating valves 37

Maintenance 40

Fixings 52

Sealants 56

Flooring 58

Bending copper tube 59

Types of system 66

Testing procedures 68

Waste systems 70

Trap seal loss 72

Traps 73

Clips, brackets, screws
    and fixings 74

PROJECTS

Project 1 **Installing a basin** 78

Project 2 **Installing a bath** 94

Project 3 **WC – Toilet cistern and pan** 104

Project 4 **Installing a cold water storage cistern (CWSC)** 116

Project 5 **Installing a kitchen sink** 126

Project 6 **Installing an outside tap** 136

Project 7 **Installing a hot water cylinder** 142

Project 8 **Removing and replacing radiators and radiator valves** 150

Glossary 158

Index 159

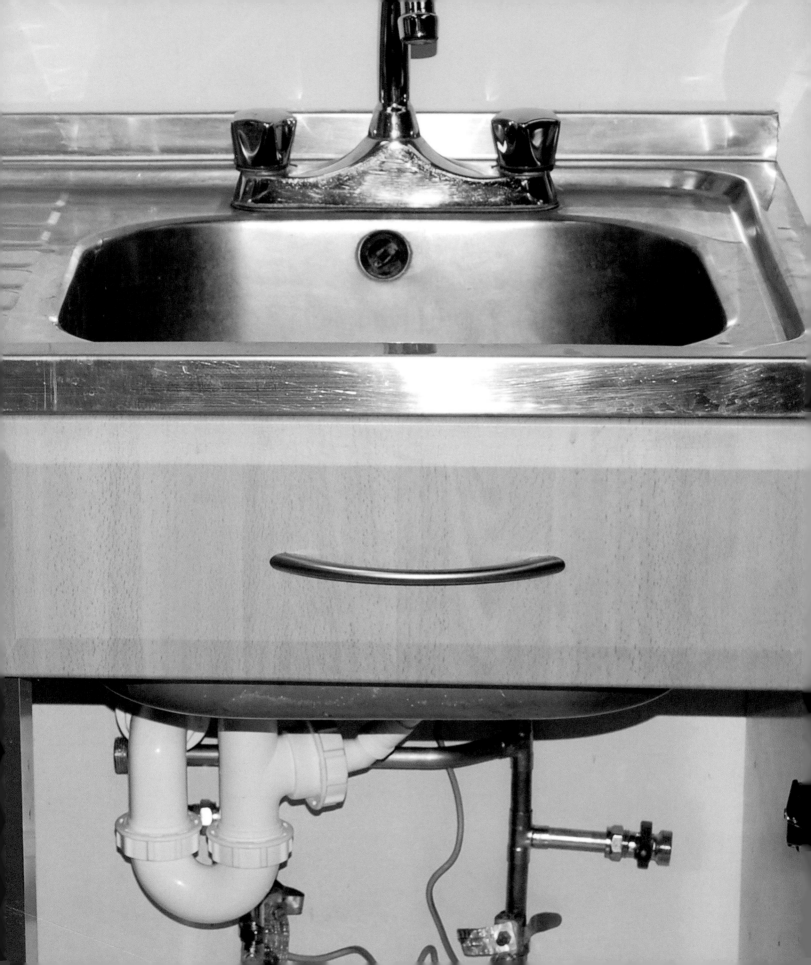

Introduction

The purpose of this book is to enable you to complete all of the projects featured here. If you have just moved into a new home, are renovating a property to resell, or have started a course on plumbing, then this book should be of great help. I have been in the plumbing industry for over twenty years and when compiling the steps for these projects my aim was to cover all the problems I have come across in the past.

The projects are presented in clear step-by-step sequences and open with a list and photographs of the tools and materials required. This recipe-style layout is a great way to undertake any of the projects, and you can always alter the recipe slightly to accommodate your own installation. Along with the projects, the book covers many aspects of plumbing, which should enable you to install new appliances and overcome most problems.

If this book has given you a taste for plumbing and you have decided that you would like to become a qualified plumber, then I recommend you enrol on a City and Guilds or NVQ plumbing course. Plumbing qualifications are obtained in two stages: level two and level three. Level two covers basic plumbing skills over 12 units including health and safety, cold/hot water to earth continuity and working relationships. The level three qualification covers domestic plumbing over seven units, including unvented hot water systems, central heating systems (including boilers and controls) and natural gas or liquid petroleum gas (LPG). This book covers some of the basic plumbing skills found in the level two qualification.

Upon completion of these qualifications and if you are over the age of 20, registration to the Institute of Plumbing and Heating Engineering (IPHE) may be sought, enabling you to become a registered plumber. Trainee registration with the IPHE is possible while you are studying at a recognised college or training centre. Whatever your intention, this book will provide you with the know-how to undertake some of the most common plumbing jobs.

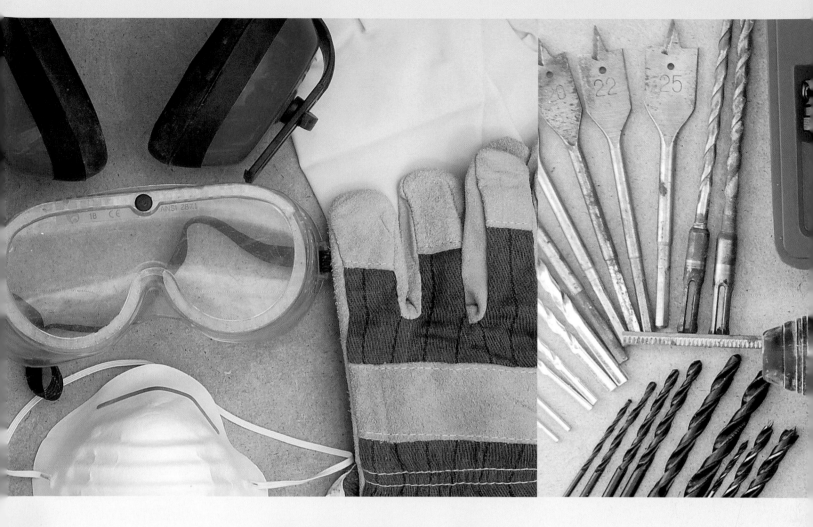

TOOLS

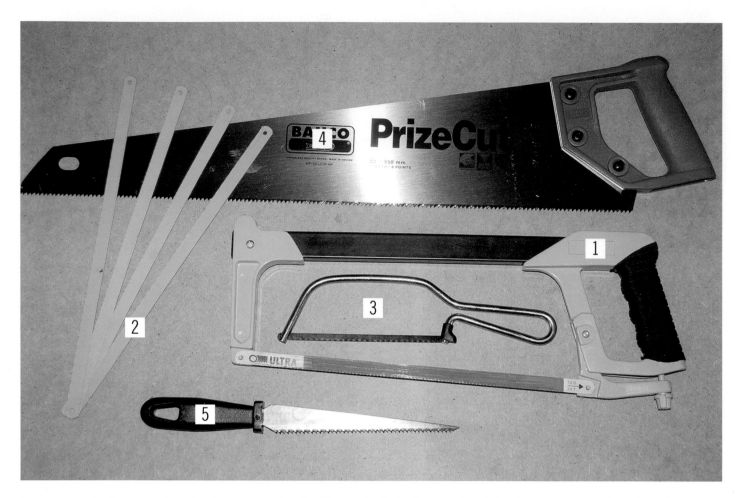

You needn't spend a fortune on tools though it is important to have a good selection at hand. The following list includes all the tools used to complete the projects in this book.

HACKSAW (1) AND BLADES (2)

A large-frame hacksaw is used for cutting plastics, copper and other pipework materials such as lead and steel. The blades on hacksaws must always be inserted with the teeth facing forward and should have 32 teeth for every 25 mm of blade for copper or plastic, and 22 teeth for every 25 mm of blade for steel and lead.

JUNIOR HACKSAW (3)

This is particularly useful for cutting in small areas where a large-frame hacksaw is too big. It is especially good at cutting off olives to remove and refit compression fittings.

HAND SAW (4)

A general use saw used in basic construction and the cutting of floorboards and joists for the installation of pipework. A smaller floorboard saw is generally used in small areas.

PAD SAW (5)

A small pointed saw used for creating holes in soft materials such as plaster board or other thin sheet materials.

MEASURING TAPE (6)

When purchasing a steel measuring tape it is advisable to select a good quality one with large numbers showing the dimensions that can be read easily in poor lighting. Tapes can sometimes get damaged, kinked or broken so select a measuring tape with a case that can easily be opened so that only the retractable tape section needs to be replaced.

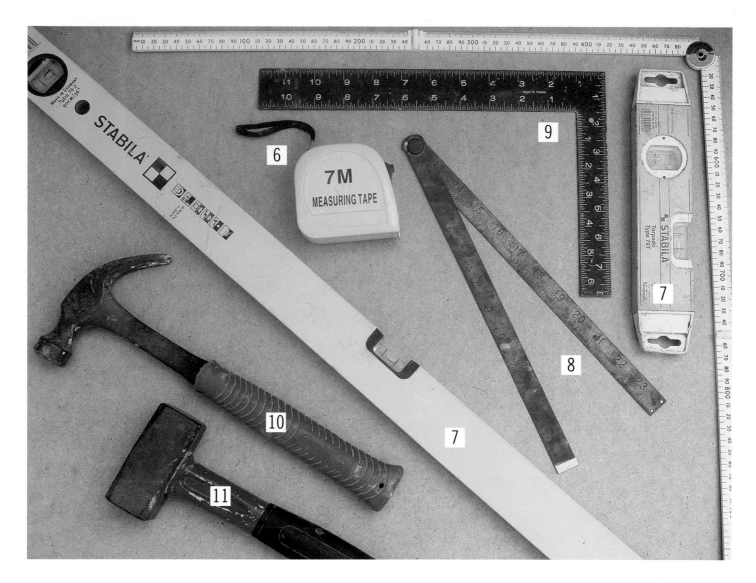

SPIRIT LEVELS (7)

A 150 mm torpedo level is very useful in small tight spaces where a large level can't be used. Some are magnetic or come with a very small 30 mm removable magnetic level incorporated within them. A standard 600 mm level is always used in every task or new project you will undertake.

FOLDING RULE (8) AND SET SQUARE (9)

A 600 mm folding rule is an essential measuring tool. The shorter varieties are useful for smaller jobs. The folding rule is handy for locating very precise bending angles (see Bending copper tube techniques on pages 59–65). Also shown is a set square, used to establish if a true 90 degree angle has been achieved. It is also used when bending copper pipes to enable precise alignment from the marked copper tube to the former.

CLAW HAMMER (10)

A basic general use claw hammer is used for all basic construction work. Combine it with a bolster to remove floorboards for the installation of pipework.

LUMP OR CLUB HAMMER (11)

This type of hammer is usually used with masonry chisels and a bolster for creating holes in brickwork for waste and supply pipework.

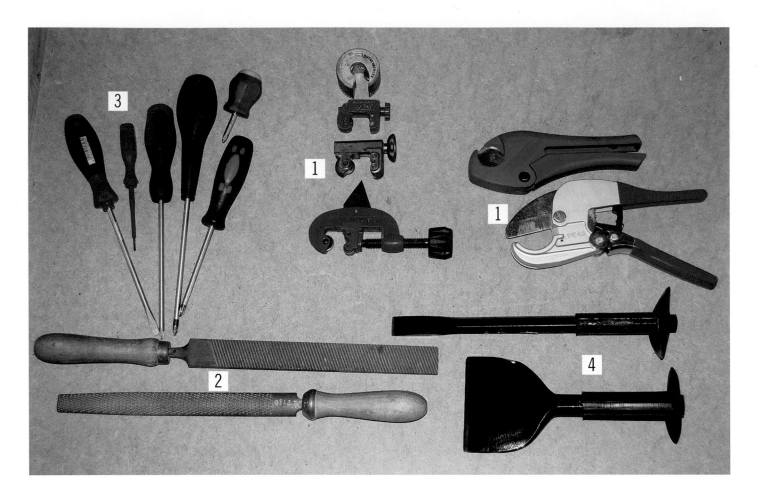

TUBE CUTTERS (1)

Tube cutters come in various sizes. Shown here is a basic 6 to 28 mm tube cutter, complete with reamer. It can be used to cut copper, chrome and plastic tube. There are smaller tube cutters that cut from 6 to 10 mm and are usually used for micro bore heating or gas pipework. Another excellent cutter is the pipe slice, which is used to cut pipes in 15, 22 and 28 mm sizes, and can cut pipe in situ and clipped to the wall.

The plastic cutters shown are a standard shear type and a ratchet type cutter. Both of these provide a straight, clean and speedy cut to plastic pipe, ensuring the proprietary insert fits neatly for a water-tight connection.

FILES (2)

Half round files and rat tail files are generally used for filing the internal part of cut tube and for removing burrs inside the pipe. The flat part of the half round file or of a hand file can be used for chamfering plastic pipes, to enable them to be connected easily and correctly. The best file for chamfering plastic pipes, especially the larger soil and vent plastic pipework, is the rasp.

SCREWDRIVERS (3)

A selection of different sizes of cross point and slotted screwdrivers are an essential part of any tool kit. Usually slotted screwdrivers are for removing old component screws and for fixing brass screws in the installation of sanitary ware. A good selection should range from small, electrical and dumpy, to large, slotted and long-reach cross point screwdrivers.

CHISELS (4)

Bolster chisels are used for lifting floorboards and cutting large holes in masonry. Cole and masonry chisels are mostly used for cutting holes for pipes to pass through masonry and brick work, especially useful when the drill hasn't made a hole large or deep enough for sleeving purposes. Wood chisels of different widths are used for notching joists and enlarging holes where timbers are in the way.

SPANNERS, WRENCHES AND GRIPS

Spanners come in a multitude of shapes, sizes and designs but for the projects in this book we will be using the 200 mm and 250 mm adjustable spanners. An open-ended flat spanner designed for 15 and 22 mm compression nuts can be an invaluable tool in reaching hard to get at nuts, such as tap connectors or compression fittings under a floor or void. A 200 mm thumb screw wrench, sometimes called 'footprints', is another very handy tool which again can be used for compression nuts in small spaces and can be easily adjusted with one hand.

Pump pliers are a type of grip that can be used for a multitude of tasks. As the name suggests they are designed to tighten and release the large nut on water pumps. They are a very versatile tool and will grip and tighten many different shapes. Stillsons are another excellent tool designed predominantly for gripping and removing LCS pipework and fittings.

BASIN WRENCHES, IMMERSION AND FLAT SPANNERS

We tend to use a selection of basin wrenches – the crow's foot wrench being the most common – and this design is manufactured in many different forms. There is also a large selection of adjustable basin wrenches in different designs and lengths. There is no one basin wrench that is the right tool for every tap installation, so you might have to use a different basin wrench for a bath than you would for a basin or kitchen sink. Small flat hand spanners can also be used to fully tighten back nuts or tap connectors. Immersion spanners are predominantly used for immersion element installations into hot water cylinders. These are designed in two styles: the box type spanner, as shown below, is designed to enable the tightening of the immersion without disturbing the foam insulation around the cylinder; and the flat type spanner, which is usually cheaper than the box type and is generally used on cylinders without foam covering, such as the older type cylinder or un-vented cylinders.

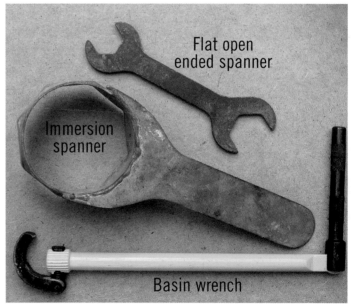

Flat open ended spanner

Immersion spanner

Basin wrench

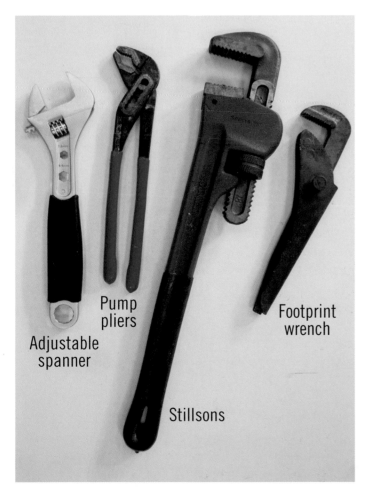

Pump pliers

Footprint wrench

Adjustable spanner

Stillsons

ELECTRIC DRILL

For the projects in this book an electric hammer drill is used. If you are drilling inside domestic premises, then 240 volts is permissible, but if drilling outside or on a building site, then the voltage must be 110 and is used in combination with a transformer. A large 24 to 32 volt rechargeable drill can also be used in any situation.

DRILL BITS

Here is a selection of drill bits required to complete all the projects in this book. These include 7, 8 and 10 mm masonry drill bits and are used with wall plugs for most applications. Fifteen, 22 and 28 mm masonry drill bits are used for creating holes in masonry for pipework and sleeving. A selection of hole saws is used for drilling cisterns and creating holes for pipework. Twenty, 25 and 32 mm flat spade drill bits are used for drilling timber floorboard or joists for pipework installation.

CORDLESS DRILL AND HOLE SAW

A large powered 24 volt cordless drill is recommended for drilling all small holes up to 22 mm. Like the electric drill it has a Special Direct System (SDS) chuck which can be converted to a basic key chuck to accept a wide variety of drills, including the hole saw. Cordless drills can be used in areas where no electric mains supply is present or in inaccessible areas when trailing electric leads could cause a hazard.

Hole saws are used to create precise holes for either the insertion of pipework through thin areas of building fabric or for the fitting of pipework connectors, such as the fittings on a cold water storage cistern (e.g. ball valve, overflow, tank connector. etc.).

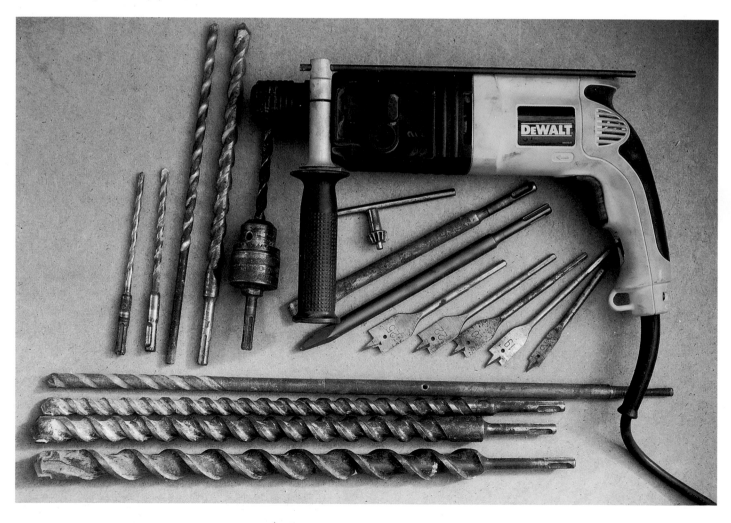

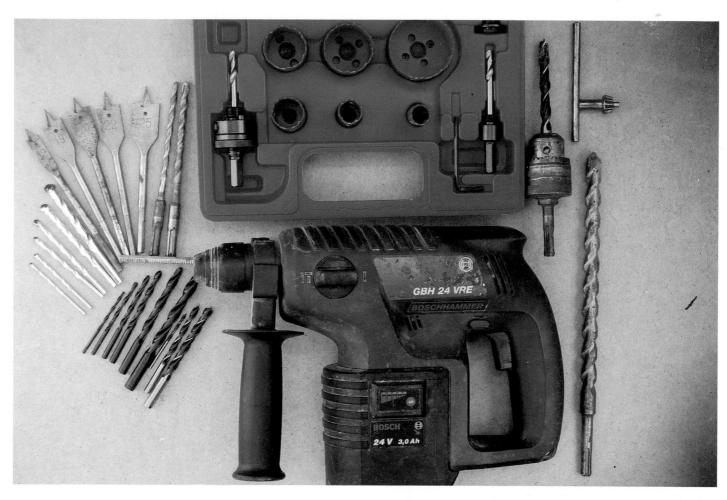

LIQUID PETROLEUM GAS (LPG) BLOWLAMP OR GAS TORCH

The liquid petroleum gas canister complete with hose and torch is generally used by the professional plumber for soldering copper pipework. The smaller hand-held models, used by both professional and DIY enthusiasts, are usually self-igniting, with some models having a cool tip to help prevent the accidental burning of yourself or the building fabric. These are used for soldering capillary fittings of either solder ring or end feed to copper tube. Heat mats are an essential part of a plumber's tool kit, and are used to ensure that no damage occurs to the fabric of the building while soldering fittings.

FIRE EXTINGUISHER

A small hand-held fire extinguisher is always a recommended addition to any DIY installation where soldering takes place and is a compulsory item for professional plumbers when working on site.

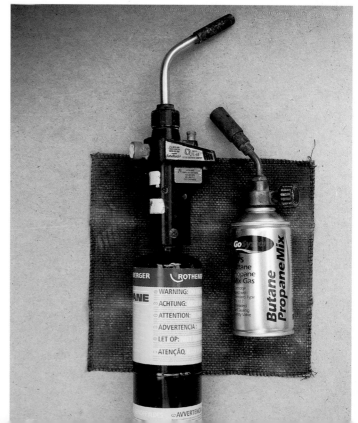

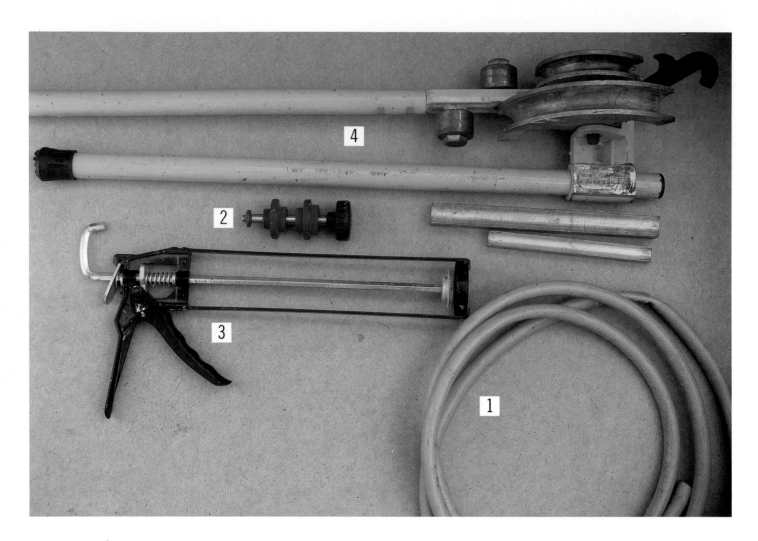

15 MM (¹/₂ IN) HOSEPIPE (1)

Generally used for draining down of pipework systems either prior to the installation of new components or general maintenance and repair. The hosepipe should be used with a jubilee clip which is essential to secure the hosepipe and prevent it from becoming disconnected from the draining point.

TAP RE-SEATER (2)

This tool is generally used in the maintenance and repair of basic screw down or rising spindle taps and valves. It is used to remove build up of scale which has developed on the seating of the washer with the help of a cutting tool screwed onto the seating of the tap.

SKELETON GUN (3)

The skeleton gun is used alongside proprietary sealants for sealing wastes and appliances.

HAND PIPE BENDER, 15 MM AND 22 MM GUIDES (4)

Hand bending machines are used to create bends and angles in copper tube. They are used in conjunction with 15 mm and 22 mm guides to ensure a smooth angle/bend is created without deformation or rippling of the tube. Larger pipe sizes require a larger stand type pipe bender.

Dust sheets

Lay down plastic dust sheets then cover them with material dust sheets to help protect work surfaces. The plastic will help contain small spillages of water, while the material dust sheet helps absorb any water and keeps the work area safe and clean. A selection of old towels or rags are good for small spills or leaks and are also used for the removal of excess flux, jointing compound and sealant.

PERSONAL PROTECTIVE EQUIPMENT (PPE)

When using a gas torch to solder in confined spaces, or where soldering may present a fire hazard or cause damage to the fabric of the building, then heat protection is required. The use of fireproof heat mats is essential and you should buy the most expensive ones you can afford. These only provide protection against the flames and heat if the torch is not aimed directly onto them as even the most expensive heat mats will burn through. There are also heat protective gels on the market that can also be used to help prevent the burning of the building fabric.

Thick gloves are another safety requirement and should be worn when installing or removing sharp metal kitchen sinks and any broken sanitary ware. Also wear them when undertaking any masonry work where chisels and hammers are used. Plastic or rubber gloves can be worn to help prevent the fluxes, jointing compounds and other compound materials coming into contact with your skin during the joining process. Safety shoes or boots should also be worn; if you are working on site these are compulsory, along with high visibility vests and hard hats.

Wear clear goggles when removing sanitary ware to protect the eyes from flying shards of metal or ceramic material. They should also be worn when drilling, along with a dust mask if drilling into masonry, wood or any other substance that produces dust. If breaking out cast-iron baths, ear defenders are very highly recommended.

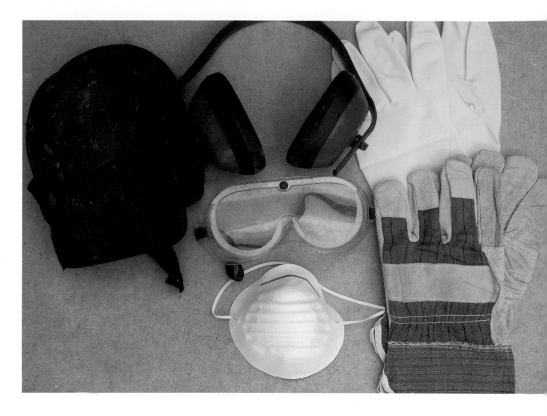

Knee pads, along with a kneeling cushion, mean you can safely and comfortably kneel without causing any damage to yourself.

Fire extinguishers are an important addition to comply with all safety regulations and they are categorised by different colour bands identifying the types of fire they can be used on. These are:

Red (contains water): used on wood, paper and fabrics.
Cream (contains foam): used on petrol, oil, fats and paint.
Black (contains carbon dioxide): used on electrical and flammable liquid fires.
Blue (contains dry powder): used on electrical equipment along with liquids and gases.

Along with the colour bands are symbols and letters denoting which class of fire they can be used for. Theses are:

Class A: fires involving solid materials – extinguished by water.
Class B: fires involving flammable liquids – extinguished by foam or carbon dioxide.
Class C: fires involving flammable gases – extinguished by dry powder.
Class D: fires involving flammable metals – extinguished by dry powder.

✘ Don't use water on electrical, oil or fat fires because of the risk of electrocution and explosion.
✘ When using carbon dioxide extinguishers, never hold the nozzle because this can cause freezer burn to your hands.
✘ Never use carbon dioxide or halon type extinguishers in confined rooms as they can cause suffocation.
✔ Always read the operating instructions on the extinguisher before use.

TECHNIQUES

EARTHING

Part 'P' regulation came into practice on 1 January 2005, and it is now a legal requirement for all work on fixed electrical installations in dwellings and associated buildings to comply with relevant standards. The relevant UK standard is BS7671:2001, 'Requirements for electrical installations' (The IEE Wiring Regulations 16th Edition). BS7671 covers requirements for design, installation, inspection, testing, verification and certification. The part 'P' qualification can be obtained by completing a short course at a recognised training centre and is graded into three sections for the professional electrician through to the plumber and kitchen fitter. Completion of this course and registration with one of the approved scheme operators would signify that you are a competent person.

The earthing of pipework and appliances is extremely important and must always be maintained. Do not attempt to undertake any electrical work yourself. All electrical work should only be carried out by a qualified competent electrician.

Equipotential bonding is the term given to the earthing of all metalwork by using conductors and earth clamps or clips, ensuring that all metalwork is kept at equal potentials preventing a build up of potentially dangerous voltages, should a fault occur on the electric system. Equipotential bonding must meet the requirements of the wiring regulations of BS7671:2001 (2004), where all metal pipework can safely be taken to earth to prevent electric shocks for anyone touching live pipework. This can also prevent the corrosion of pipework.

The diagram on the right shows a typical earthing arrangement, with the main equipotential bonding conductors connected to the metal water main, which should be connected as close to the point of entry as possible. The metal gas main should be connected to the consumer side of the meter and be a maximum distance of 600 mm (23½ in) from the meter.

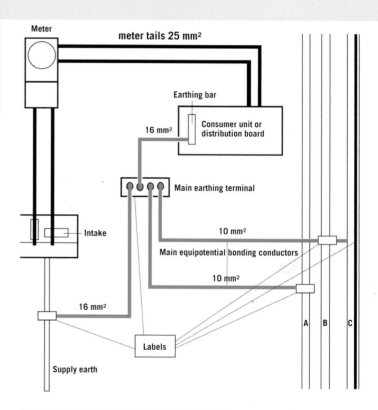

Key
A = metal water pipe; B = metal gas pipe; C = other extraneous conductive parts.
a) The main equipotential bonding conductors may be separate (as shown) or looped with unbroken conductors.
b) Local electricity distribution network conditions may require larger conductors.
c) Labels = safety electrical connections – DO NOT REMOVE
d) The green lines are earth wire

SUPPLEMENTARY BONDING

Supplementary bonding is the earthing of other exposed metal parts within the hot and cold water and the heating systems. The metal parts such as pipework, radiators kitchen sinks, baths and metal wastes are typical examples of where supplementary bonding is required. These areas may have become unprotected and isolated from the earth by plastic fittings or cisterns for example, and now require supplementary bonding to maintain the conductivity of the pipework to earth.

A typical layout of supplementary cross bonding is shown in the diagram below. The diagram is divided into zones, with each one representing the degree of risk/proximity to water services. The supplementary earth bonding wiring and connectors are shown in green.

TEMPORARY BONDING

If you are repairing or cutting into an existing pipework installation you must provide earth continuity. An example of this is if you were cutting into a cold water main to connect a supply for an additional appliance, then as soon as you have cut and separated the pipe, earth continuity is lost and the metal parts above this cut will now not be earthed. To overcome this problem a temporary bonding wire is used. This consists of a length of 10 mm^2 earthing conductor with a 250 V rating minimum, and crocodile clips fitted to each end. This temporary bonding wire will be suitable for pipe sizes up to 28 mm.

The temporary bonding wire is attached to the pipe before cutting to bridge the gap you are going to make and the wire is only removed after the job is completed.

EARTHING STRAP

The earthing strap is easily connected to the pipework, by simply wrapping the thin metal strap around the pipe and threading it through the main body. It should then be pulled tight before tightening the screw, ensuring the retaining nut is securely tightened.

Care must be taken to ensure it is firmly connected to the pipe and the earthing tags notice is connected and facing forward and covering the cable termination. If the pipe is painted, then it must be cleaned off back to the copper before connecting the earth strap.

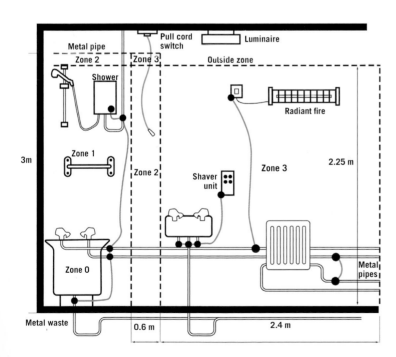

SAFETY AND PREPARATION

The following information should be adhered to closely to ensure that each job is carried out safely and according to best practice.

■ During the preparation it is best to construct a method statement of how you plan to approach the job or project in hand. The time you set aside for this varies from task to task, but I recommend on average one hour. It is a good idea to compile a comprehensive list of tools and materials needed for each job. Although this can seem like a waste of energy, it will save you time and effort in the complete installation.

■ For all types of work undertaken, a selection of both plastic and material dust covers must be used. The plastic sheeting should completely cover the floor area followed by the material sheet on top. Even if you are only changing a tap washer you should always lay down a dust sheet before placing your tools onto the floor and cover all surfaces around the work area. Occasionally, partial lifting of the carpet may be necessary if you are removing a radiator or installing new pipework under a floor.

■ If you need to use a gas torch, make sure you have adequate protection in the form of a very good quality heatproof mat and possibly a heat protective gel. Never solder where the flame can be drawn into a hole. An example of this is when soldering pipework close to a hole in a ceiling, floor or duct work. The flame can be drawn into the void or loft space and can cause dry timbers or other combustible material to smoulder and ignite, causing a fire. If you have to make the connection close to any hole it is better to use either a compression or push fit fitting.

■ If you are carrying out work where access to a kitchen or bathroom cupboard is needed, ask the customer to clear out the space, removing any cleaning agents or other materials around the work area which may be flammable.

■ Ensure all services are fully accessible and are in full working order. Many problems occur when starting an installation because of inoperative or faulty valves. Once you have ascertained the working order of the valves and ensured that the correct valve isolates your work area, close the valve as much as possible and drain the pipework to empty.

■ If you are working on a hot or cold water service, open all taps above and below the work area, ensure toilet cisterns have been flushed and shower and bidet valves are fully open as this can sometimes be overlooked when draining down. An example of this could be that if someone flushes the WC after you've cut into the same supply pipe at a lower level, capillary attraction will be lost and water will flood into your work area. This is why your tool kit should always include a bucket and a small container to help catch any residue of water which may be present in the pipework.

■ The health and safety of yourself and the people around you must always be your first concern, whether you are changing a washer or installing a completely new plumbing system. Ensure all the tools you are using are fit for purpose. Tools that are not up to standard could include hammers with split handles or loose heads, files with broken or missing handles, damaged heads to screwdrivers, mushroomed heads to chisels, broken or missing cutter blades and adjustable spanners that don't adjust. Always inspect every tool before use.

■ When lifting heavy objects keep your feet slightly apart, bend your knees, keep your arms close to your body and your back straight with your chin tucked. Ensure a tight grip with your hands and not just your fingers. Lift by straightening your legs. Although this may seem like a basic technique it is just too easy to hurt your back if you don't follow the correct procedure.

PIPE MATERIALS

It is important to establish the material you are going to use or join to. If you are working on an old property it is likely that you will have to connect to either lead pipe, copper or galvanised iron.

If any of the pipework has been plastic coated or has been painted a yellow ochre, then this is probably carrying a gas supply and should be avoided, but don't rely on this sort of identification alone; always fully examine and investigate the pipework you are going to work on before carrying out any work. When removing a section of metal pipework for either repair or to enable additional branches to be added, it is essential that the earthing continuity is maintained. This is achieved by connecting a temporary bonding wire across the gap which should be connected prior to the removal of any pipework.

LEAD PIPE

Lead pipe is easily recognised as a dull grey pipe that can be marked very easily and is usually laid in long lengths. This type of pipework is very often found in old properties. If lead pipework is present in the property and it is impossible to completely remove it, then you will have to connect your new pipework to it. Lead used to be traditionally connected by either wiped or rolled joints. These connections were constructed by experienced plumbers and are quite difficult to make. Because lead pipe is no longer used as a supply pipework in today's properties these types of connection are no longer used. The method now adopted to connect lead pipe to approved types of pipework is with a compression fitting, which shouldn't be disassembled like a normal compression fitting, but pushed onto the lead pipe and tightened by using two adjustable spanners or grips. No jointing compound is used on the lead connection side of the fitting. If the other side is connecting to copper, then normal compression techniques are used.

LOW CARBON STEEL PIPE

Although LCS (low carbon steel) pipe looks very similar in colour to lead it is far stronger and the fittings are threaded onto shorter lengths of pipe. If connecting a water supply to an existing LCS water main using the same material, then only galvanised low carbon steel can be used. But if the connection is to heating pipework then untreated LCS is acceptable.

Both these connections are made by removing a threaded fitting from the LCS and screwing the new section of pipework into the fitting. To install LCS pipework requires a threading machine of either the hand held or stand type.

Copper tube should not be connected to LCS heating systems because of its corrosion properties when mixed with steel (electrolytic action). If you have to cut the LCS in place and it is not possible to thread the pipe to make a connection, then you can use a fitting which is very similar to the lead connection, and is installed in the same way. These connections enable you to connect to either LCS, copper or plastic. Copper and plastic can also be connected to LCS from a threaded fitting with either a male or female connector.

COPPER TUBE

Copper tube is distinguished by its copper colour which sometimes has a green patina. Copper tube is available in four grades: X, Z, Y, and W.

Grade X is generally used in most domestic installations with pipe diameters from ranging from 12 to 54 mm. Half hard straight lengths are available from 1 to 6 m and are also available in a chromium finish, which is a more desirable finish for exposed installations in bathroom, kitchen or radiator connection pipework. Grade X pipework can be connected by either capillary or compression fittings, but should not be used underground.

Grade Z has a thinner wall than grade X, and the pipe is made stronger by being hardened during the manufacturing process. Like grade X it is available in 1, 2, 3, and 6 m hard straight lengths and in a chromium finish for 12 to 54 mm diameter pipes. Because of its thinner wall, grade Z cannot be bent, but can be installed underground.

Grade Y has a wall thickness of 1 mm making it the thickest grade of the four, and is generally used in underground installations. It is fully annealed which means it is soft copper and has diameters of 15 and 22 mm. It is supplied in soft lengths of 25 m and can easily be formed into wide angle bends by hand without the use of machine bending equipment. It can be supplied with a coloured plastic coating in either blue for water services or yellow for gas services.

Grade W is the last of the copper grades. This small bore pipework is also fully annealed with diameters of 4, 5, 6, 8, 10 and 12 mm, and like grade Y can easily be formed into wide angle bends by hand without the use of machine bending equipment. For tighter and shorter offsets and bends a small hand bending machine is available. The pipe is available in soft coil lengths of 10 to 30 m and is generally used in micro-bore heating and short gas supplies. Like grade Y it can be supplied in a coloured plastic coating – yellow for gas and white for heating.

Where pipework is to be installed in solid floors internally, then a plastic coating with air channels can be used to help improve its thermal insulation properties.

PLASTIC PIPEWORK

Plastic pipework is generally very easily recognisable and can be laid in very long lengths. It can be manufactured of polythene, polyethylene, polypropylene, UPVC (unplasticised polyvinyl chloride) or ABS (acrylonitrile butadiene styrene) and can only be connected in certain ways. The first three plastics are generally colour-coded blue and are used for underground mains water supplies, hot water and hot water heating, with all of them using compression fittings for the joining or connection process and only polythene and polypropylene also using the fusion welding process. Polypropylene pipe can also use a push fit 'O' ring connection.

UPVC and ABS can use compression, solvent weld or push fit 'O' ring connections, as explained in Joining methods (see pages 31–36). All these plastic systems can be connected to copper.

Copper tube

Plastic pipework

FITTINGS

There are many different types of fitting available in today's market. These range from traditional soldered, compression and threaded fittings, which require specific tools to use, to the push fit plastic or copper type which require a simple 'push fit' as the name implies.

COPPER FITTINGS

Copper fittings come in various sizes from 4 to 54 mm tubes in domestic properties, with the basic sizes for most domestic installations usually ranging from 10 to 28 mm. A general selection of fittings is shown in Picture 1.

Left to right, clockwise: coupling; bend; push fit bend; tee.

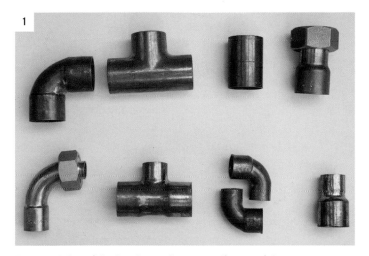

Top row, left to right: bend; equal tee; coupling; straight tap connector. Bottom row, left to right: bent tap connector; 22 x 22 x 15 tee; two street elbows; internal reducer.

All these fitting are of the end feed type but are available in solder ring also, and Picture 2 shows a tee and bend which are of the solder ring type, along with a push fit bend and coupling.

Most of these fittings are self explanatory, with the exception of the street elbow. This is a bend that has a socket on one end and a spigot on the other, enabling it to be connected to other fittings without the need for additional pipework, and creating a close connection.

The internal reducer is simply inserted into a fitting to reduce the pipework, whereas the 22 x 22 x 15 tee has a reduced branch at the top of the tee. When purchasing

tees they are always ordered by what size the connections are through the tee and then the branch. For instance, the tee shown here is ordered as 22 x 22 x 15. Equal tees, are fittings with the same size connections all round.

22 x 22 x 15 tee.

Equal tee.

BRASS FITTINGS

Picture 3 shows a selection of typical brass compression fittings.

Top row, left to right: bend; straight coupling; back plate elbow.
Bottom row, left to right: 22 x 22 x 15 tee; tank connector;
½" to 15 mm copper to iron straight.

When connecting LCS (low carbon steel), similar fittings are available, but if you are cutting into a section of pipework and because it will always have two fixed ends, the need for a union fitting is often required. Picture 4 shows a selection of LCS fittings.

Top row, left to right: bend; equal tee; coupling; 135° male.
Bottom row, left to right: ¾" x ¾" x ½" tee; union, 135°.

The soldering and compression fitting connection techniques are described in Joining methods (see pages 31–36).

PLASTIC FITTINGS

Plastic connections are generally manufactured to the same designs as copper fittings. Picture 7 shows a selection:

These are typical push fit and glue connections, and the techniques for installation and removal are covered in Joining methods (see pages 31–36).

Top row, left to right: bend; equal tee; reducer.
Bottom row, left to right: bend; reducer (glued rigid plastic fittings); large water main bend.

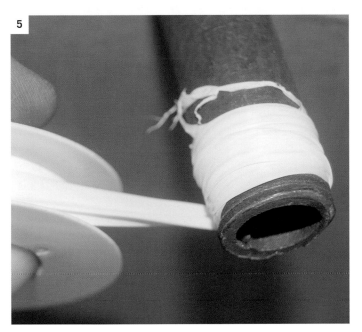

5

The thread of a low carbon steel pipe is being wrapped with PTFE.

When connecting LCS pipework onto a fitting, the thread must firstly be wrapped with PTFE (polytetrafloraethelene), more commonly known as 'plumber's tape for everything', as in Picture 5! This is a thread sealing tape. Once the pipe thread has been fully wrapped the fitting

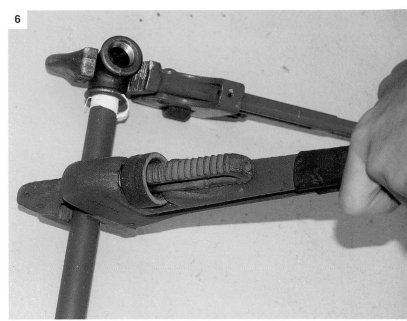

6

The tightening procedure for LCS using two pairs of stillsons.

can be wound by hand onto the pipe, and using a combination of either two pairs of stillsons (as in Picture 6) or one pair of stillsons and a pipe vice, the fitting is then tightened onto the fitting.

Picture 8 shows a selection of larger waste fittings which can be either push fit or solvent welded (glued).

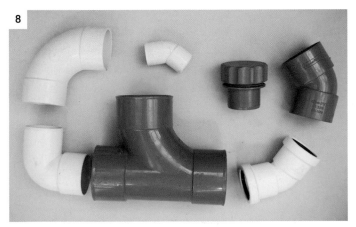

8

Top row, left to right: bend; 135° angle; access plug; 135° angle.
Bottom row, left to right: tight street bend; equal tee; 135° angle.

FUSION WELDING

Fusion welding is mainly used on water and gas services and is used in conjunction with specialised heat welding equipment. This type of joint uses a specially designed tool which heats a metal wire inside the fitting by means of an electric charge. This wire is located just below the internal surface of the fitting and when heated melts and fuses the fitting and pipework together.

The connection of compression, push fit and solvent welded connection techniques are described in Joining methods (see pages 31–36).

CUTTING PIPEWORK

Many different tools can be used for cutting pipework but the hacksaw is the only one that can be used for all types of pipework.

CUTTING COPPER TUBE

Copper tube can be cut using many different types of cutter. The standard tube cutter comes in a range of sizes and can cut copper tube ranging from 4 to 54 mm in diameter.

1

2

1 The standard tube cutter has two rollers on one side and a blade on the other. The pipe is placed onto the rollers and the handle is then tightened until the blade touches the pipe.

2 With one hand holding the tube, revolve the cutter around the tube, slowly increasing the pressure by tightening the rollers until the copper cuts and separates.

3 When the pipe is cut, ream the newly cut end ensuring that it remains at the same diameter as before the cut. This is achieved by using a reamer, which is a V-shaped section of metal located at either the end, or in a folded area to the side of the tube cutter. Rotate this into the cut end to ensure the full bore of pipe is maintained and the pipe is clean of any debris or particles that may have occurred during cutting.

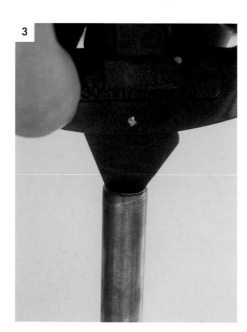

3

The same method is used with a mini cutter, but because of its size it does not come with a reamer, so one must be used from a standard tube cutter. The mini cutter is ideal for cutting small pipework in situ and can expand to cut pipe with a diameter of up to 22 mm.

4 Another very popular cutting tool is the pipe slice, which comes in three sizes: 15, 22 and 28 mm. Pipe slices are easier to use than standard tube cutters and pipework can be cut in situ where pipework is clipped to the wall or in an inaccessible area, such as under a basin or in a cupboard.

4

Push the pipe slice onto the pipe and turn it towards the cutter ensuring the rollers and the cutter are firmly pushed towards the copper tube at all times. If you try to turn the cutter in the opposite direction it will not cut properly and will release itself from its hold. This is an excellent cutting tool but because it is round, it tends to roll away from you towards the nearest open floor or hole!

5 When cutting copper pipe with a hacksaw care must be taken that it is cut squarely. This can be achieved by wrapping masking tape around the pipe on the mark, ensuring the tape goes around the pipe in a complete circle. Alternatively, place a pipe clip onto the pipe and use it as a guide.

The teeth on the blade should face forward and there should be 34 teeth for every 25 mm of blade. Use the full length of the blade when cutting. Once the pipe has been cut through completely file the inside of the pipe with a small round file. Remove any roughness and smooth off the edges around the cut end of the tube with a flat file. This ensures that the tube diameter remains the same and is able to connect easily into its fitting. The same method is used for chromium-plated pipework.

When cutting LCS (low carbon steel) a hand tube cutter similar to the one used for copper tube can be used. The same techniques are also used when cutting with a hacksaw with the exception that

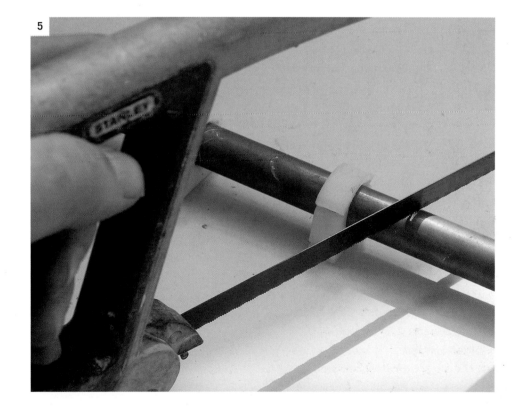

5

instead of holding the pipe by hand it is held in a pipe vice. The hacksaw, however, should have 22 teeth per 25 mm. LCS is too strong to be hand reamed and so has to be filed.

CUTTING PLASTIC PIPE

The cutting of plastic pipe is generally best made by means of purpose-made cutters; these can be of either the shear type or of the ratchet design. These cutters are best used on plastic supply pipework because they ensure a straight cut which is essential for achieving a watertight connection.

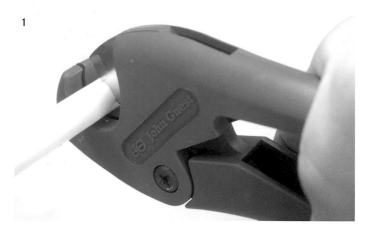

1

2

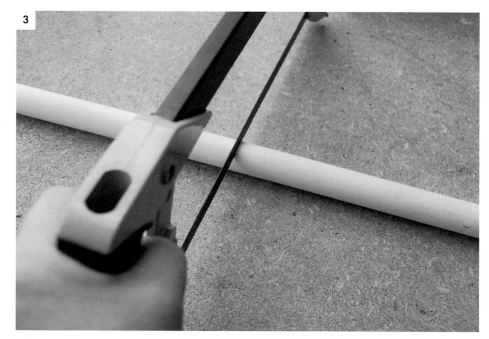

3

Another method of cutting LCS is to use a large stand-alone threading machine, which is capable of not only threading pipework but also cutting and reaming. The pipe is locked centrally in the machine and then cut with a large roller-and-blade type cutter, which once removed is reamed by a centrally set reamer. The cutter and reamer are held in a fixed position and the pipe is turned by the machine.

The smaller blue cutters (1) are designed to cut plastic pipework up to 22 mm, and are generally used on domestic installations. They are light and easy to use and require little effort to create a straight cut in plastic pipe.

The larger ratchet-type cutter (2) is a more professional tool designed for plastic and vinyl pipework. This is a much more robust tool, slightly heavier than the previous cutter but again easy to use and will easily cut through the strongest of plastic / vinyl pipe to a much larger 50 mm diameter size.

If you have to use a hacksaw (3), then the same procedures used for cutting copper pipe apply. A square cut is essential and can be achieved with the help of masking tape or a pipe clip as described earlier for copper.

Once the plastic waste has been cut and the inside and edge have been de-burred and filed clean, the edge should be given a slight chamfer with a file or rasp to enable a secure connection into the required fitting.

JOINING METHODS

Joining can be achieved in a number of ways, using a variety of fittings and tubes. The first method use capillary attraction and is a recognised tried and tested procedure of soldering copper tube. Once a good join is made it will last for many years and is still the preferred joining process for many large contractors. The other tried and tested method is the compression fitting which can be easily fitted, but can develop slight water leaks. The third process covered is the push fit connection which is in plastic and copper fittings.

JOINING COPPER TUBE

Push fit fittings are used in the same way as plastic push fit fittings. The finished appearances of these fittings are very neat and can be used in areas where it is not advisable to use a naked flame. There are the three basic methods of joining copper tube: soldering, compression and copper push-fit.

1 The first task is to ensure that the end of the copper tube is clean and bright and this can be achieved by rubbing with a wire wool pad, an emery cloth, a specially designed wire brush or a scouring pad. Care must be taken when separating wire wool strands as they can easily cut your hands and it is always best to use scissors to cut and separate the wire wool into small pad-like sizes. Gloves should be worn when separating wire wool, but they don't need to be worn when cleaning copper tube.

Once the tube is clean the next step is to decide what sort of solder should be used. If the pipework used is for wholesome (drinking and washing) water, then lead-free solder MUST be used. For other uses, such as sealed central heating systems, a lead solder is suitable.

2 The choice then is whether to use either an end feed or solder ring fitting. An end feed fitting requires an additional supply of solder and a solder ring has, as the name suggests, an integral ring of solder, which makes an extremely neat joint. If using this type of joint, you must be certain that it is of the lead-free type if used on potable installations. This is certified by a BS and lead-free mark on the side of the fitting. These fittings are much more expensive than the end feed type.

3 Once the pipe and the fitting have been cleaned the next process is to flux both the pipe (3a) and the fitting (3b) using a small brush. The use of flux is very important in the cleaning and helps in capillary attraction.

4 Insert the adjoining pipe into the fitting and ensure it is fitted to its maximum depth (4a), then, using a damp cloth, wipe off all the excess flux (4b). Failing to do this will make the solder run around and over the joint making a very untidy joint and if soldering in a vertical position the solder can run down the pipe and bear a slight resemblance to candle wax.

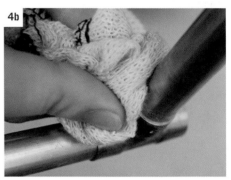

5 Now the pipe has been cleaned and fluxed and assembled with any excess flux removed. All that remains is the addition of heat from the gas gun along with a small amount of solder (5a). When joining a 15 mm pipe, a 15 mm length of solder is required for each joint made. This system is used for 22 mm and 28 mm pipes alike. These are generally the size of supply pipes used in domestic premises today.

Angle the solder to approximately the correct length of solder (i.e. 15 mm) and use the hottest part of the flame, which is just above the internal cool flame. This is shown in the inset photo (5b) by the insertion of a match into the cool area of the blue flame without igniting it.

Plumber's tip
When soldering, the point above the cool area must be used onto the fitting and solder for the solder to melt and achieve a fast connection. If the torch is too close and the internal blue core is used, then the fitting can take too long to heat and the solder will not melt effectively. This can also result in the fitting becoming black and dirty, which means that the solder will not be able to adhere to the copper and possibly create a leaking joint.

6 Using the correct procedure, apply the solder gently to the joint. As the solder melts, capillary attraction draws the solder into the joint completely, filling the void and making the connection between pipe and fitting, thus creating a very water-tight joint (6a). Using capillary attraction means you can solder in any position, even upside down! The solder only needs to be fed into any one area of the joint, so if the fitting and tube are in an inaccessible area, we can be

6a

6b

reasonably certain that the solder will run all around the fitting with the aid of capillary attraction. When making a

joint using a solder ring fitting (6b), then the same process is applied, except no additional solder is required.

7 When connecting copper tube using compression fittings, the only additional jointing material used is a potable/wholesome approved jointing compound. The first step is to dismantle the fitting into nuts and olives. If connecting to an old imperial size pipework or ¾ in and above, then the correct size imperial olive should be used, but for ½ in pipework a regular 15 mm olive is acceptable.

7

8 First, place the nut over the pipe followed by a correct size olive (8a). Apply a small amount of jointing compound with a small brush. Ensure that you have applied enough all the way around the pipe and over the area between the olive and the end of the pipe (8b).

Insert the pipe all the way into the fitting, then slide the nut along the pipe, pushing the olive onto the fitting (8c). Tighten first by hand, taking care that the nut is fitted squarely onto the fitting to ensure cross threading doesn't take place. Then, with the use of a combination of adjustable spanners, pump pliers or open-ended spanners, tighten the fitting with one tool holding the centre of the fitting and the other turning the nut clockwise, in turn crushing the olive onto the pipe to make a

8a

8b

8c

8d

water-tight joint (8d). This non-manipulative compression type of joint should not be used on buried or hidden pipework and should only be used in areas that are accessible.

Plumber's tip

There are currently a variety of push fit fittings for copper on the market. Here are just a couple of them which are designed to have the appearance of either solder or compression fittings.

The first fitting shown is of a simple push fit design (1). The fitting is simply pushed onto the pipe to create a connection (2). Gripper rings inside the fitting hold the pipe firmly while an 'O' ring type connection prevents the water from leaking, making a very neat joint. But unlike compression or end feed, once fitted they cannot be resealed or repaired. If there is a leak, they can only be cut out and replaced.

The other type of push fit fitting shown is a larger style (3) and can be easily removed with the use of a specific tool supplied with the fitting. The same method for connecting pipework into the fitting is used, and removal is simple: first clip the removal tool onto the pipe, then pull towards the fitting to release it (4).

This particular fitting can also be used to connect plastic pipework to copper with the addition of a copper insert fitted into the plastic. Compression and push fit fittings are an excellent connection where water is still present in the pipe and when soldering is not an easy option

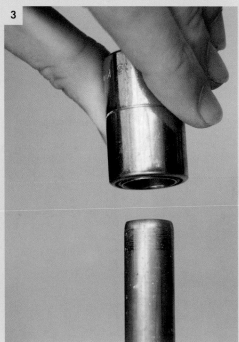

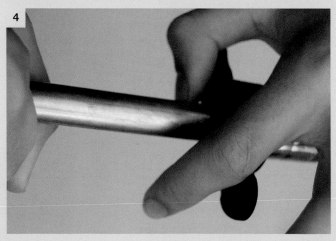

JOINING PLASTIC TUBE

Plastic pipe and fittings are making a big impact on today's plumbing systems. New builds and refurbishments use a large amount of plastics in both hot and cold water services, as well as in central and under-floor heating systems.

There are now many types and varieties of plastic pipes and fittings. On numerous new developments or refurbishments plastic pipe is used as the 'first fix' – this is pipework that may be run under flooring or in hidden box work and terminates with short tails into rooms for hot and cold services or radiator connections. Copper is often taken from this plastic tail for the 'second fix' of pipework connections that will be on view. One of the reasons plastic pipe is preferable to other more rigid materials in first fix, is that it is distributed in long, flexible coiled lengths and is easily fed through holes and voids.

1 When using plastic pipework it is imperative that you cut squarely and cleanly, as described in the Cutting pipework (see pages 28–30). An insert is then pushed into the end (1a) before the fitting is securely fitted onto the pipe (1b). This type, like many others, can also be used to connect copper, by pushing the pipe into the fitting in the same manner (1c). Removal is just as simple with the end ring pushed towards the fitting to release it (1d).

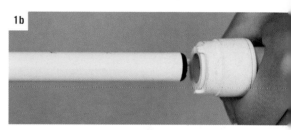
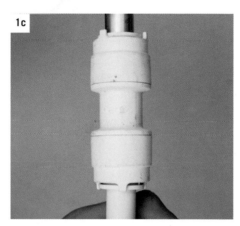
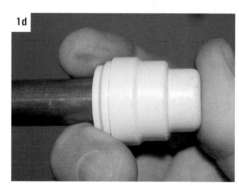
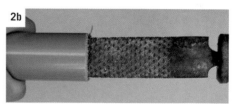

2 When connecting plastic solvent waste pipe, make sure the pipe is cut clean and squarely. The pipe edge is then angled (2a) and de-burred by using either a coarse file or rasp (2b). It is then cleaned using wire wool and a proprietary cleaner (2c).

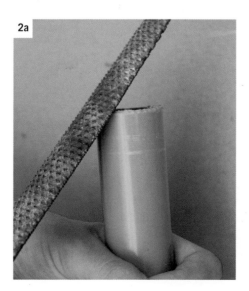

3 Adhesive is then applied to the inside of the fitting (3a) and the outside of the pipe (3b). When using any solvent glue or cleaner always refer to the manufacturer's instructions and use in a well ventilated area. Insert the pipe into the fitting and slightly turn to engage the glue to firmly adhere the fitting and pipe. Remove any excess glue with a damp cloth. The fitting should then be left to fully set according to the manufacturer's instructions.

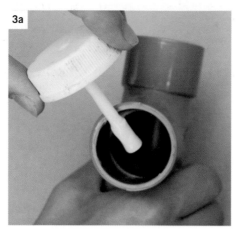
3a

3b

4 Two alternative methods for connecting waste pipework are the push fit system of fittings generally only used internally. The pipe ends should be slightly angled and de-burred after cutting, as explained in Step 2. A silicone lubricant is then applied to the pipe end with a cloth (4a) to enable the rubber connector to slide easily onto the fitting (4b).

4a

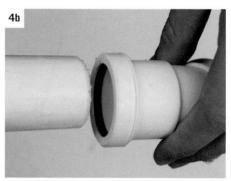
4b

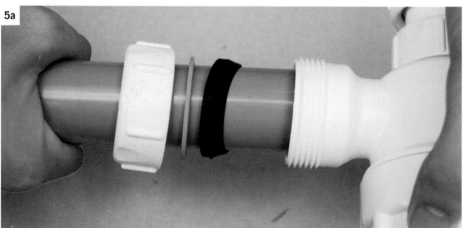
5a

5 The other connection is the compression fitting used in a similar method as a copper fitting only no jointing compound is necessary. The fitting is dismantled and the plastic nut is placed onto the pipe (5a), followed by the blue rigid thin plastic washer and finally a rubber olive, which must be placed so that the bevelled edge faces the fitting and the flat edge faces the rigid plastic washer. The nut is then firmly tightened by hand (5b).

5b

ISOLATING VALVES

Isolating the water services is an important process of any project you undertake, and locating the water mains stopcock and other isolating valves, including down service gate valves, is very important before attempting any work. You need to be sure you know the location of the hot and cold water isolation valves and that they are easily accessible and in good working order.

If the property you're working on is a flat or maisonette, then the water mains stopcock should be located just inside the property, with perhaps another situated next to the water meter if one is installed, which would be located in a cupboard in a communal area. If the property you're working on is a house or bungalow then the main stopcock could be situated outside and close to the boundary wall, where it may be located under a metal covering with a hinged lid. On a lot of older properties the lid gets removed and either a stone slab gets laid over it or if it is a newly paved area, block/slab paving which can be easily removed for access is fitted. Underneath the lid or slab is a chamber set in the ground to a depth of approximately 75 cm to 1.35 m, at the bottom of which is a stopcock. This can only be turned by means of a long crutch type device called a stopcock key (1). There will also be another stopcock outside your boundary but this is the sole property of the water authority and should not be tampered with (2).

Along with the external stopcock the property should have an internal stopcock, and it may be sited just inside your property (3). On early properties the stopcock was normally sited by the main entrance to the property, with the floorboards having not been fixed and finger holes drilled in them for easy removal. Once the boards are lifted a stopcock, which was very often connected to a lead main, can be clearly visible on the dirt floor. This design was used for many years in properties with

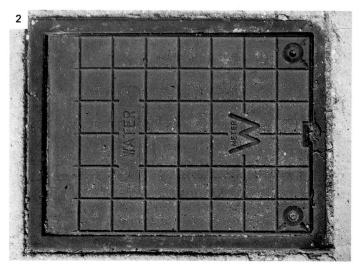

A water authority stopcock cover.

suspended flooring. But stopcocks can be found almost anywhere, from kitchen to bathrooms, so very often if you are working on an unfamiliar property some time must be allowed to find its exact location.

Wherever you find the stopcock, care must be taken when turning the valve for the first time as they can often break if forced too hard, through

Internal stopcock.

lack of use and corrosion. The chances are that even if the stopcock turns easily, the gland nut will leak (the repair of gland nuts is covered in Maintenance (see pages 40–51). But if the stopcock does break it is

Stopcock key.

Stopcock and drain-off.

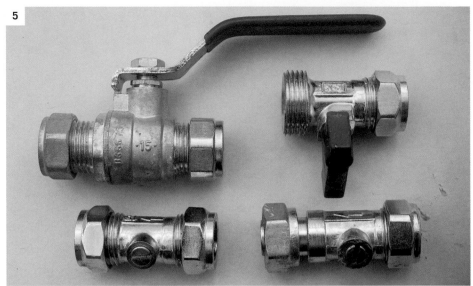

Left to right, clockwise: lever type valve; washing machine valve; standard isolating valve; tap connector valve.

generally in the fully open or fully closed position. If it is located outside the only alternative is to dig down to the valve and replace it. Whether the valve is inside or outside it may entail asking the water authority to turn off the water supply so a repair can be carried out. This can sometimes result in other properties close to you having their water also being turned off if you have a shared water main. In the case of flats or maisonettes, it is very possible the whole of the block may have to be isolated. Alternatively, the pipework can be frozen (see page 39).

In new build properties the recommendation is that the water mains stopcock should be sited inside the property as close to the main entrance as possible and should be easily accessible. It should only enter a maximum of 100 mm into the property before a stopcock is fitted with a drain-off cock connected above it (a combined stopcock and drain-off is also acceptable) **(4)**.

Individual stopcocks or quarter turn isolating valves **(5)** are usually found under kitchen sinks and sanitary ware pipework, along with the rising main connection to the ball valve in the cold water storage cistern.

Gate valves **(6)** are generally found in either the loft area and are connected to the down service pipework from the cold water storage cistern or in the airing/cylinder cupboard sited on the cold feed to the hot water cylinder. Be careful when turning of any gate valves and ensure that they only turn off the correct pipework. In an airing cupboard the gate valve, which is connected to the cold feed of the cylinder (the lowest point on the cylinder), should be turned to isolate the hot water supply. Any other gate valve sited in this cupboard could be connected to the heating system and should be avoided.

Gate valve.

When turning gate valves always be prepared for them to break. If they haven't been used for a long time and are of poor quality, then they may very possibly break so with this in mind come prepared with a spare gate valve and rubber bung to stop the flow of water from the cistern.

Loosening of the gland nuts is a recommendation prior to turning the central gate valve or stopcock. This will enable the valves to move more freely than before. You may get a slight weep of water from the gland nut once loosened, but this can be easily repaired as explained in Maintenance (see pages 40–51). The other valves that should work to full capacity are the ball valves, which are generally situated in cold water storage cisterns or WCs.

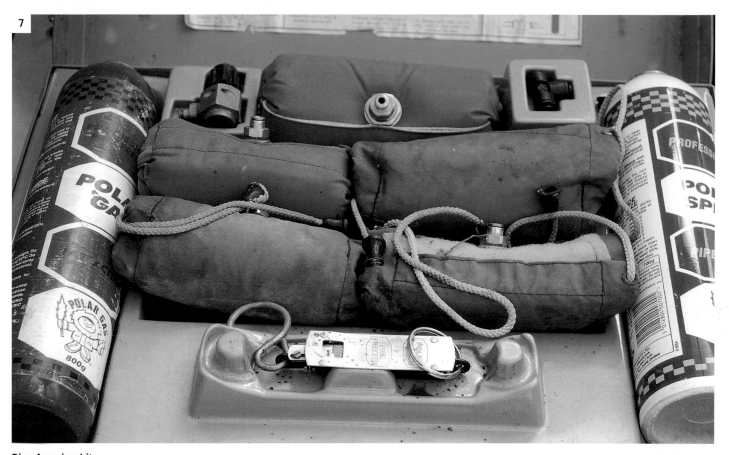

7

Pipe freezing kit.

If the pipework cannot be isolated because of a faulty valve or for repair/replacement of a valve, then the pipework can be frozen. This can be done in two ways, the first of which with the use of a small aerosol type canister and muffler. The muffler is wrapped around the pipe and each end is tightened with plastic ties. A small tube is connected from the freezer can to the muffler. The can is then discharged into the muffler and after the recommended waiting time the pipe can be worked on for a set time. When using this type of freezing kit always follow manufacturer's instructions.

A larger, more professional gas system is shown **(7)**. This kit uses mufflers with ties. It is connected in the same way but the amount of gas used is measured with scales. More gas is used for longer diameter pipe **(8)**. The pipe is only frozen for a set time and the manufacturer's instructions should be followed.

Professional electric and gas freezing are alternative method of freezing pipework, but should only be carried out by professional pipe freezers.

8

Freezing 15 mm copper tube.

MAINTENANCE

The maintenance of plumbing systems is a very big part of plumbing and it is essential to maintain all appliances so they are in full working order. A tap or ball valve for example left dripping and unattended does not only waste water but has the additional problems of staining and possibly creating water damage to the property.

STOPCOCKS

The maintenance and repair procedures for stopcocks, gate valves and isolating valves are quite simple. The main problem with stopcocks and gate valves is leaking gland nuts. These types of valves have a rising spindle which means that as the valve is opened the spindle and head rise allowing the water to flow. After a period of time the packing which prevents water from leaking around the spindle deteriorates.

To repair this problem turn the valve off and release the gland nut and slide it up the spindle **(1)**. Wrap PTFE sealing tape around the spindle between the gland nut and the valve **(2)**, then using a small screwdriver carefully push the thread sealing tape into the valve **(3)**. Once this has been achieved replace the gland nut and tighten with an adjustable spanner **(4)**. Open the valve and test. You may find that more packing is required, if so repeat the procedure until it is fully repaired. After the repair the valve will probably be far tighter than before; this is

normal and will not affect its operation.

If a stopcock is not turning off fully and water is letting by (leaking), then the water board should be called to fully isolate the supply. The other method is to freeze the pipework, as described on page 39. Once you have isolated the valve, and using an adjustable spanner **(5)** release the main body from the base of the valve **(6)**. On the base of the body is a jumper connection which holds the washer. If the problem is a faulty washer, remove it and replace it with a new one of the same size (the size of the washer depends on the size of the valve: ½ in, ¾ in, etc.) **(7)**.

If the problem appears to be not with the washer but with the seating of the washer onto the valve, then use a tap reseating tool with the correct type cutter and screw the reseating tool into the base. Turn the cutting disc onto the seating of the washer to ensure a firm connection between washer and valve **(8)**. Refit the body back into the valve and test.

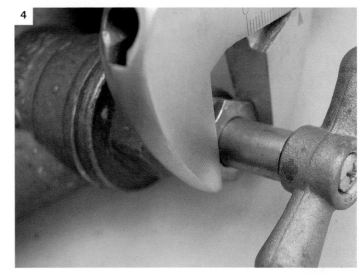

GATE VALVES

Gate valves are very different. Although they are of the rising spindle type they do not use washers; instead they isolate the water by means of a full way brass gate, hence the name. These types of valve should only be fitted to low-pressure systems, usually down services from cold water storage cisterns. If a gate valve is not fully closing or letting by, then once you have isolated the faulty valve with the use of an adjustable spanner **(1)**, remove the main body from the valve **(2)**. You may be able to clean and remove any debris or scale which has accumulated in the gate way section of the base or gate. If once reassembled it still remains faulty then replacement is necessary.

QUARTER TURN ISOLATING VALVES

With the basic quarter turn isolating valves generally used to isolate individual taps or ball valves, no maintenance is required. Sometimes with this type of valve a leak develops through the middle screw adjuster which turns the ball. If this occurs then replacement is the only option.

RADIATOR VALVES

With radiator valves the main problem is again usually leaking gland nuts on the rising spindles of lock shield valves. To repair them, follow the same repacking procedure used for stopcock valves but make sure that the heating system is turned off. Both radiator valves to the radiator must be turned off even if only one valve is leaking. Suitable floor protection should be placed under and around the valve and containers should be at hand to collect any water which may leak from the valve. Once you remove the gland nut from the valve **(3)**, the water that is still present in the radiator may start to leak out of the top of the gland nut. If there appears to be a lot of water the radiator should be drained down as described in the Removing and replacing radiators and radiator valves project (see pages 150–157). The gland nut can then be repacked, reassembled and tested with the heating on and off.

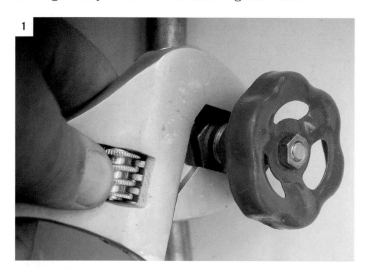

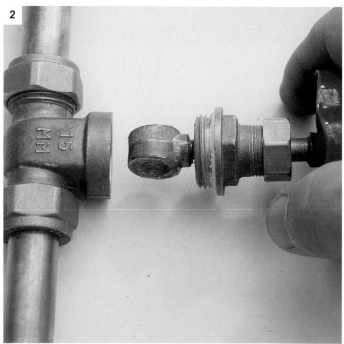

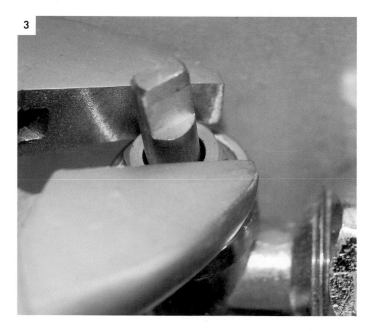

TAPS

Before caring out any maintenance or repairs to taps, the plug should always be fitted to prevent any screws or small components going down the plug hole. To maintain basic rising spindle taps such as cross headed pillar taps often found in older bathroom and kitchen installations, the techniques used are the same as described in the stopcock section. These include repairing the gland nuts, replacing washers and reseating.

The only difference being that on some taps there is a shroud covering the gland and body nuts. This shroud is either chrome or plastic and either screwed or push fit depending on whether the tap is old or modern. To remove the shroud completely the cross head has to be removed. This is achieved by either removing a small screw or grub screw from the side of the head or removing the hot and cold identification tops. Once the shroud is either lifted or removed the same procedures are used.

Screw down bib cocks and garden taps should be treated as stopcocks. Taps that have non rising spindles generally have a large covering acrylic or metal head, which after the water has been isolated, has to be removed by removing the centre disc with a small bladed screwdriver or similar tool, and undoing the screw **(4)**. The head can then be removed by pulling it off the spline connector on the body of the tap. Once the head has been removed the tap body can be released with a pair of adjustable spanners **(5)**. Care must be taken not to move the tap from its fixed position as this will create a leak below the appliance. Assuming the problem was a dripping tap, remove the tap body and then the washer and replace it with a new one **(6)**.

4

5

6

TIP

If the taps seems loose when removing the body of the tap to replace the washer, wrap a soft cloth around the spout to help prevent any damage and hold the spout still with a suitable pair of grips. Once the washer has been replaced don't forget to tighten the tap from beneath the basin as described in the Installing a basin project (see pages 78–93).

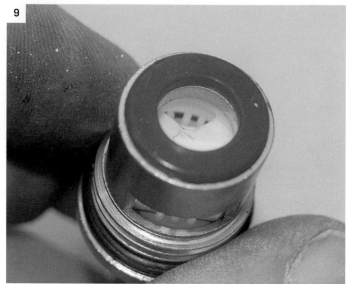

'O' ring and ceramic disc replacement

If the problem was water leaking around the top of the tap, then with the body removed, remove the top circlip and put somewhere safe to refit it later **(7)**. Tap the spline end gently removing the threaded section away from the body **(8)**. There should be one or two 'O' rings fitted to the top of this screwed section. These are easily replaced with new 'O' rings before reassembling the tap. If the tap requires reseating then the same procedure as described in the stopcock section is followed. Once the tap has been repaired, test and examine tap connections in case of possible disturbance during repair.

Quarter turn ceramic disc taps are very popular with many new modern installations and they work on the principle of two discs, one turning with the handle and the other fixed below it. When the holes in the discs are in line, then the water is released. There is not much maintenance involved in these taps but if they begin to constantly drip, then the ceramic disc section will

probably need replacing **(9 and 10)**. The removal of the tap body is very similar to other taps with the exception of the removal of the tap handle. The removal of these differs from tap to tap, but there is often a small grub-like screw that retains the handle.

Mixer taps are generally a combination of the taps described above, and are maintained in the same ways. With kitchen mixer taps, there is a common problem with the centre spout leaking from the swivel joint, this can be easily repaired by removing the small screw (possibly a grub screw) from the rear of the spout. If the screw is at the front of the tap, then it is very possible they have been fitted backwards. This now means the hot and cold has reversed places in the spout with the hot running along the outside skin, which makes the spout very hot to the touch when being moved. Once the screw has been removed the spout should lift up from the mixer. Remove and replace the defective 'O' rings, then refit and test.

Ball valves

The maintenance procedure for float operated valves or ball valves is generally associated with noisy or leaking problems. When dealing with leaky ball valves causing the overflow pipe to be activated in either the storage or toilet cisterns, the first course of action is to isolate the water supply to the ball valve; this valve should be either the quarter turn or stopcock type and should be located close to the cistern. Check the float and arm first ensuring that the float has not got a build up of scale

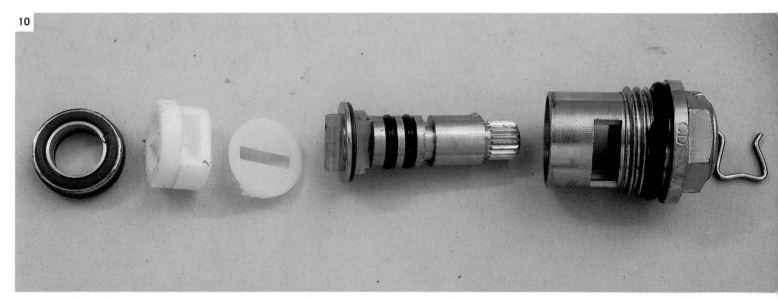

or a hole which would prevent it from rising in the water and shutting the valve. Test the arm for ease of movement **(11)**. A build up of scale, especially in the heating feed and expansion (F & E) cistern, is often a problem. With the F & E heating cistern the valve is often fixed in the closed position, due to the very small amount of water used in the heating system.

Once you have tested both the float and arm the next part to look at is either the diaphragm or washer depending on the type of ball valve installed. All ball valves must comply with the British standard 1212 (parts 1–4). The Portsmouth ball valve (part 1) is not used widely in new installations today, but is still found in many properties. These types of ball valves use a piston type

action when the float and arm rise in the water, pushing a rubber washer onto a plastic valve preventing the flow of water. To repair the washer in this type of valve requires you firstly to release the large nut **(12)** that holds the main valve assembly. Remove the fibre washer and replace it if defective. Then remove the plastic valve and ensure it is clean and free of debris or scale. Remove the split pin

TIP
When cleaning out the valve section of a ball valve always discard the old washer and replace it with a new one, as disturbing the old connection and reusing the existing washer could result in a leaking connection.

13

(13) that retains the arm, and then remove the internal chamber holding the rubber washer (14). Replace the washer (15) and reassemble the ball valve. Reconnect the valve ensuring the fibre washer and plastic valve are connected. Reinstate the water and test.

Diaphragm type ball valves (parts 2–3) are manufactured in either brass or plastic, and unlike the Portsmouth ball valve has a top water outlet. The same procedure in testing the float and arm (16) is used along with an additional adjustment which can be made to either the screw attachment on the float, or the screw attachment located on the arm and valve assembly (16a).

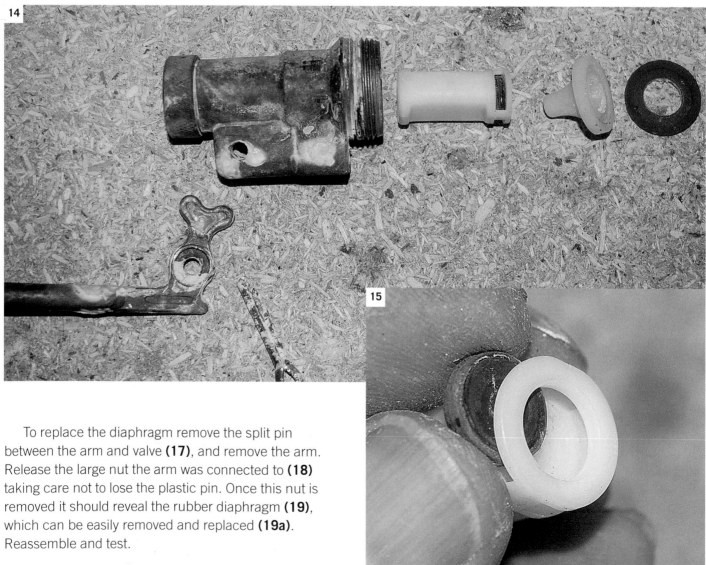

14

15

To replace the diaphragm remove the split pin between the arm and valve (17), and remove the arm. Release the large nut the arm was connected to (18) taking care not to lose the plastic pin. Once this nut is removed it should reveal the rubber diaphragm (19), which can be easily removed and replaced (19a). Reassemble and test.

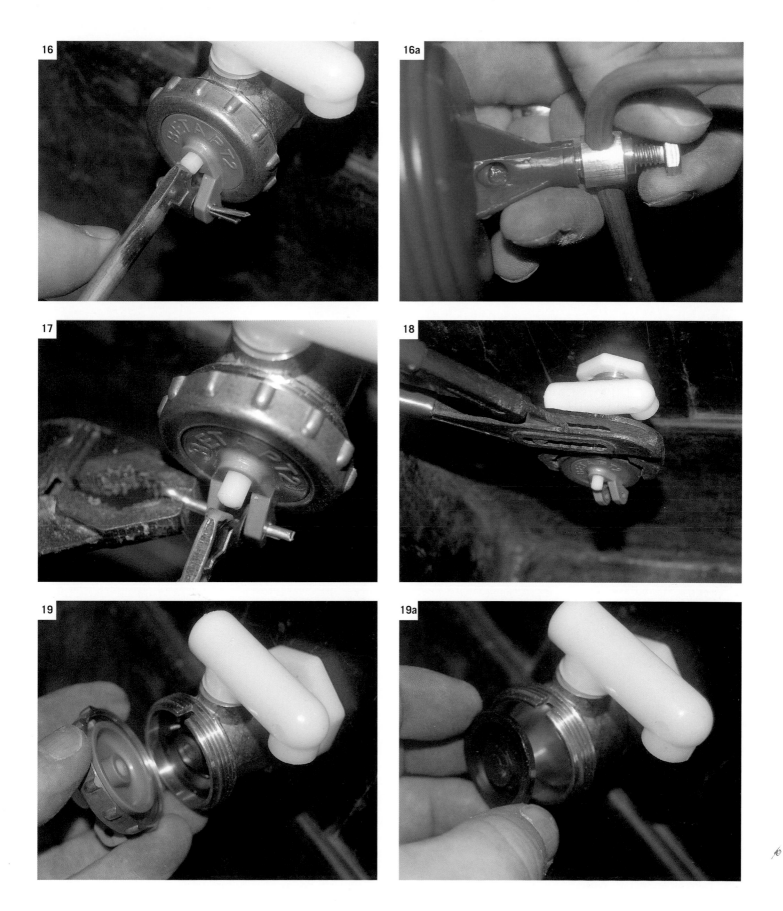

COLD WATER STORAGE CISTERNS (CWSC)

If water enters the cistern too slowly, then the problem could be with the plastic valve, which would entail removing the nut that holds the complete ball valve **(1)**. The plastic valve can then be cleaned of scale or debris and replaced ensuring the fibre washer is fitted in place **(2a, 2b, 2c)**. The plastic valves in ball valves are designed to have either a large outlet for low pressure supplies (this could be a connection to a toilet cistern from a CWSC), or small outlet for a high pressure supply (fitted to either direct cold water systems or onto storage cisterns).

If there is a hammering noise in the pipework every time a tap is opened or the toilet is flushed, then a possible cause is a high pressure supply acting upon a low pressure ball valve. A simple replacement of the ball valve for an equilibrium ball valve can be fitted to resolve this problem. The installations of ball valves are covered on pages 116–125. Another possible noise in pipework comes either from pipework that is not securely clipped or that is fitted too close to another surface, such as timber.

Another problem associated with cold water storage cisterns are the overflow connections. These should always be examined for leaks between the cistern and connector and if an insect mesh is fitted check that it is clear of obstructions **(3a and 3b)**. Also check the entire length of the overflow pipe for any possible leaks or damage to the pipe. Inspect the lid of the cistern

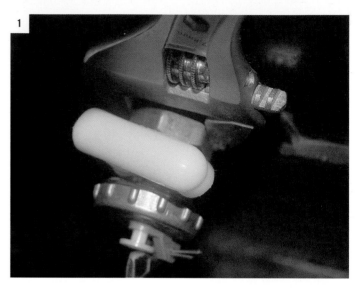

1

2a

2b

2c

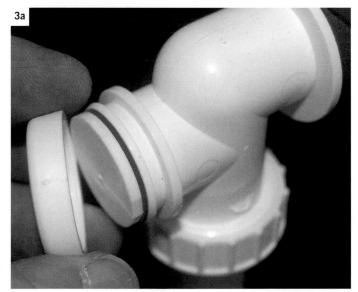

3a

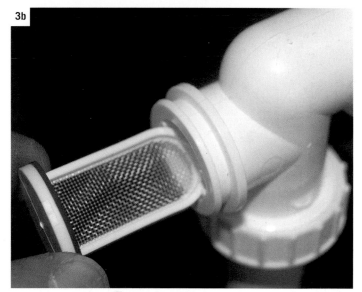

3b

ensuring that it is tightly fitted and if it has been fitted to current bylaw 30 standards, check the insect mesh on the vent along with the grommet around the vent pipe if present. It is very important to have a close-fitted lid on the cold water storage cistern; wasps' nests, pieces of wood, roof tiles and even pigeons can end up festering in

there and contaminating the water supply.

Once the cistern and overflow has been fully checked, the supply pipework and down service pipework should be examined for leaks. Check that both the pipework and cistern are fully protected with insulation against possible freezing.

TOILET CISTERNS

To remove either of the siphons on the low level or close coupled cisterns, refer to the installations procedures for the project WC – Toilet cistern and pan (see pages 104–115) and simply reverse the procedure.

With toilet cisterns, the overflow should be inspected for leaks between the cistern and the overflow pipe, and also along the pipe's full length until it terminates outside the building. Poor flushing is a basic problem with toilet cisterns. This is generally caused by a torn diaphragm in the siphon. If the cistern is of the low level type, then once the water has been isolated from the cistern and all the water has been removed it is a simple task to remove the siphon. This is achieved by releasing the plastic nut which connects the flush pipe to the cistern and removing the flush pipe. Then remove the large plastic nut from the bottom of the cistern. With this nut removed the siphon can then be unattached from the handle

assembly and removed from the cistern **(1)**. Remove the metal hook from the diaphragm rod **(2)** and then remove the rod and plastic diaphragm assembly from the siphon **(3 and 4)**. Replace the plastic diaphragm, reassemble and test.

If the cistern is of the close coupled type, then the same process is used to remove the siphon and diaphragm after the cistern has been removed from the pan by removing the two retaining bolts between the cistern and the pan and released from the wall by undoing the retaining screws. Replacing the diaphragm on both these types of cistern can take time, with the close coupled taking the longest. However there is a siphon which is manufactured in two parts which allows the diaphragm section to be removed through the releasing of a single nut on the siphon. The two-part siphon is the only design where removal can be carried out without isolating or removing the water.

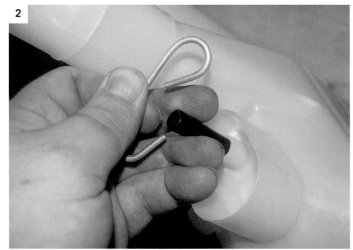

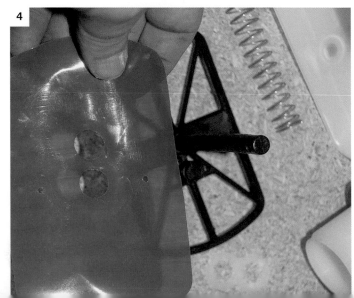

TRAPS

Before removing any of the components of a trap it is advisable to wear suitable rubber gloves because the silt water from traps has a strong bacterial smell which is difficult to get rid of once on the skin. It is also a good idea to fully protect the area directly around the trap and to place a bucket under it.

Place the plug into the waste before you remove the trap, this stops any water from running through the waste if somebody accidentally turns the tap on. Remove the trap by carefully undoing the top plastic nut which is connected to the waste outlet. This can be done either by hand, with pump pliers or grips. Release the compression nut from the waste pipe and remove the trap **(1 and 2)**. Empty the contents of the trap into the bucket, clean the trap and refit it. Remove the plug and test.

These procedures are also used for the maintenance of 'P', 'S', running or washing machine traps. For the removal of bottle traps the same preparation is used but the bottle portion of the trap need only be removed for cleaning purposes **(3 and 4)**.

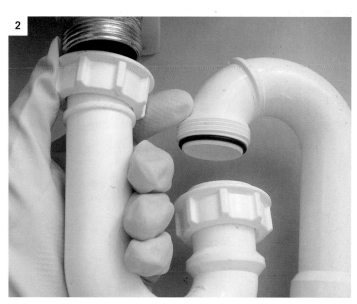

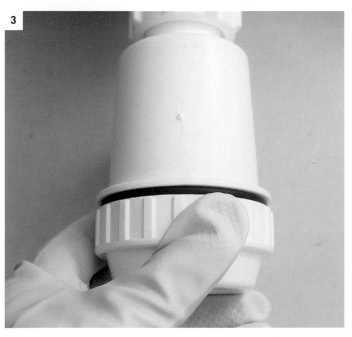

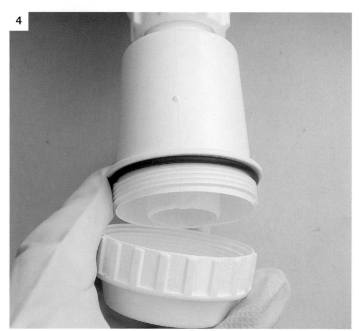

FIXINGS

When fixing into any surface care must be taken not to penetrate pipes or electrical wiring which could be present. Prior to drilling or fixing the use of a pipe and wire detector is advisable.

There is a variety of fixings currently available on the market. Screws are generally sold stating their length in inches and then the gauge or width of the shank. Generally, the gauge can be converted to millimetres as follows: 6=3mm, 8=4mm, 10=5 mm. All the sizes given here are for guidance purposes only and the size of the screws depends on the structure and situation of the installation.

For fixing clips, steel countersunk screws of either the crosshead, pozidrive or slotted type are generally used. These range from ½ in x 8 to 1½ in x 8, with the smaller size used to fix saddle clips.

If fixing clips or brackets in external areas or where moisture could occur, then brass or alloy screws must be

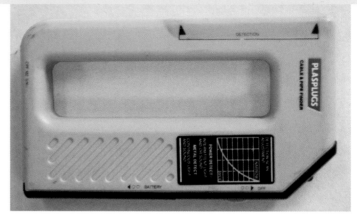

Pipe and wire detector.

Left to right: a selection of cavity fixings. Nylon wall plug (masonry); spring toggle; hollow wall anchor; metal screw fixing.

used. Examples of this are when installing sanitary ware such as a WC where brass screws are used to fix the WC to the floor. When fixing either waste or rainwater pipework brass or alloy screws should also be used. These could be either round head, raised head or counter sunk.

When fixing radiator brackets 2½ in x 10 or 12 are typical screw sizes which may be used, but as previously stated each installation is different and there is no definitive screw size. A number of radiators used today come supplied with fixings and plugs, which can sometimes have a nut type head, which has to be tightened either with a socket or grips.

If fixing into a timber surface use basic steel screws; if the surface is chipboard the use of chipboard screws is advisable. When fixing onto a brick or masonry surface you must first drill a hole and insert a plastic wall plug before a screw can be used.

The size of the screw gauge you use determines the size of the masonry drill and the colour of the plastic wall plug. A screw size gauge of 6–8 requires a yellow wall plug and a 5 mm masonry drill. These are quite small drill and plug sizes and are generally used for small screws, possibly saddle clips. A screw gauge of 8–10 requires a red wall plug and a drill size of 6 mm. This is also a small sized plug and screw and is generally used for light work only. Brown wall plugs are used for most general applications. They require a 7–8 mm masonry drill which will accept 8–14 screw size gauges. The next size is the blue plug which has a masonry drill size of 10 mm and a 14–18 screw gauge. These are usually used for large, heavy radiators, heavy sanitary ware and boilers.

When drilling holes to take these plugs, care must be taken to ensure that the depth of the hole is not too deep. There are many different types of cavity wall fixings, and opposite is a selection of some of the most commonly used **(1)**.

Plasterboard and cavity wall fixings can either be plastic wall plugs, various toggle connectors, or a screw type which screws straight into the plasterboard. The plastic wall plug type is designed with small plastic fins on the end to prevent it from pulling back through the surface and is generally used for lightweight support only. This nylon wall plug can be used in aerated blockwork as well as general masonry (2).

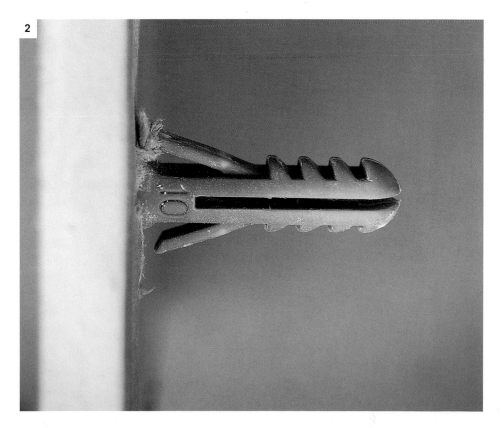

2

The toggle type connector comes in various forms. The first one shown is the threaded bolt and spring toggle, which is installed by first drilling a hole large enough to take the folded toggle. The toggle, complete with threaded bolt, is then inserted into the hole complete with item that you are connecting to the surface – like a radiator bracket for example (3). The spring toggle can then open and the bolt tightened (4). The only drawback is that the bolt can never be removed without the spring toggle being lost in the cavity.

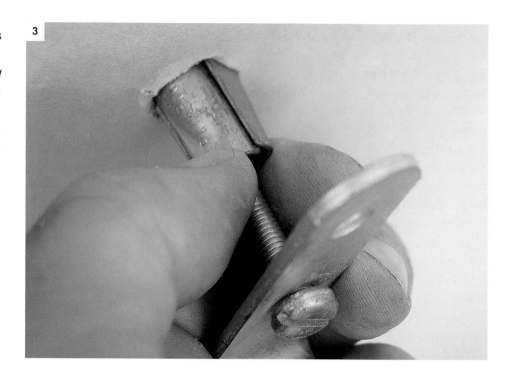

A selection of steel, floorboard, masonry, and galvanised clout headed nails are also used in basic plumbing installations with steel nails used in basic construction such as CWSC and hot water cylinder bases to boxing in pipework and fitting bath panels. Floorboard nails, as the name suggests, are used for fixing flooring back after pipework installation. The designs of these nails help prevent the floorboard from splitting when nailed. Masonry nails are used when fixing to brickwork, and galvanised nails are generally used to prevent them from corrosion where moisture is present.

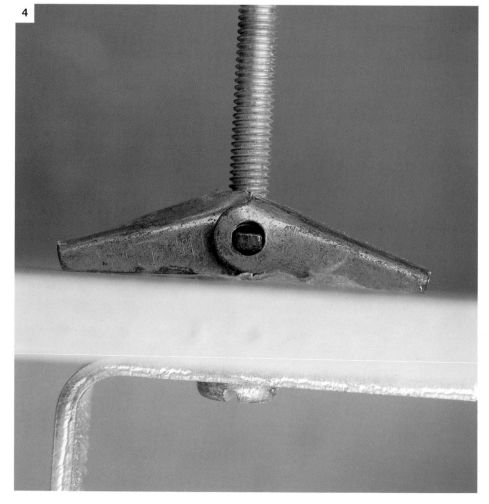

The other form of cavity fixing is the hollow wall anchor, which is also fixed by drilling a hole in the wall and then fully inserting the toggle sleeve and threaded bolt into the hole (5). With the use of a specially designed ratchet style tool (6) the bolt is pulled away from its sleeve causing the sleeve to be crushed onto the inside of the cavity creating a strong connection (7). The threaded bolt of this connection, unlike the spring toggle type, can be removed without losing its internal connection. Both these cavity fixings can be used for light to heavy support depending on the length of fixing.

Plasterboard and cavity fixings of any type can't be relied upon to support very heavy items, and if unsure always use additional support. The screw type plasterboard connection is made simply by just screwing straight into the plasterboard (8 and 9). A screw supplied with this fixing is then screwed directly into the fixing. This is a general fixing used for light support such as pipe clips.

SEALANTS

The correct sealing material is very important to prevent contamination of potable/wholesome water. The terms potable and wholesome refer to water suitable for human consumption. The making of water-tight connections is a very important aspect of plumbing, and care and consideration should be taken when making these to ensure the correct material is used.

THREAD SEALING

The old method of sealing pipework threads was to apply an oil-based jointing compound and then wind a length of spun hemp around the thread. This was an excellent method of thread sealing but this technique is now only used on low carbon steel heating installations, and must never be used on any hot and cold (potable/wholesome) water systems.

The techniques now used in domestic hot and cold water installations are to either apply a potable/wholesome jointing compound over the thread or to use PTFE thread sealing tape around the thread. Jointing compound and PTFE should never be used together.

SEALING FITTINGS

When sealing fittings such as taps or ball valves to tap connectors, a fibre washer is used to seal between the two. These are available in 15 and 22 mm and should be supplied with the tap connector. The sealing of cistern tank connectors should only be made with plastic or rubber washers and no sealant or jointing compound should be used.

When fitting wastes and overflows into sanitary ware or sinks, the use of either plumber's putty or silicone sealant can be used as both have excellent sealing properties. Plumber's putty and silicone sealant can also be used to seal around tap holes, prior to fitting the taps, to help prevent water running back from the spout and leaking through the tap hole.

Silicone sealant can also be used in the sealing of appliances to the wall or tiled area. All the techniques for using these sealants are described in the Joining methods (see pages 31–36)and the Installing a basin project (see pages 78–93).

260° C (500° F), and along with water can also be used on steam, natural gas, LPG, kerosene, air and refrigerants.

PTFE

Can be used in a multitude of ways from sealing the thread on fittings to the repacking of gland nuts on taps and valves. It can be purchased in either single or double wrap (double is twice the thickness of single wrap) reels or a special PTFE designed for gas.

PLUMBER'S PUTTY

Plumber's putty is a waterproof non-setting putty specifically designed for sinks, waste water fittings, joints in sanitary ware and soil pipes. It is a permanently flexible sealant which can be over-painted and will seal metal, ceramic, enamel and PVC.

JOINTING COMPOUND

The particular potable/wholesome jointing compound shown can be used to seal threads of fittings up to 75 mm (3 in) on copper, brass, iron polyethylene and PVC. It has a working temperature of 0° C (32° F) to

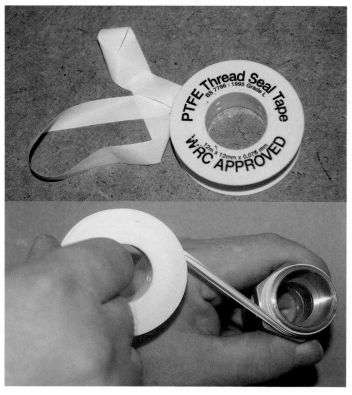

FLOORING

When installing pipework in properties with floorboards, it is essential to lift the flooring and drill or notch out joists to run new pipework. Before starting any work, an inspection should be made to check for pipework or electrical wiring. This should be done with a pipe and electric wire detector.

When lifting floorboards, it is always best if you can lift them in full lengths. If the floorboards are of a standard type, then place a bolster in the gap along one side close to one end and, using a hammer, prise up the board slightly. Repeat this procedure on the opposite side and continue this along the whole length of the board, lifting each section up and clearing it of the joists (1).

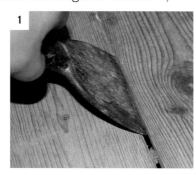

When the board has been completely removed, turn it over and tap out all the floorboard nails, mark the board with a number to help replacement later and move it to a safe place. Remove any nails which may have remained in the joists during lifting.

If you only require a short length of board to be lifted, then the same procedure is used until you reach the joist section which will allow you enough room to carry out your job. Lift the board past this point and, using either a hammer or bolster, wedge it under the board and cut through the board using a floorboard or hand saw close to the nails ensuring that when both sections are replaced they are still supported by the joist (2).

If the flooring is tongued and grooved then the tongue will have to be cut out first by either using a specialised tool designed for the job or a sharp pad saw. If, however, the flooring is chipboard and it is not possible to lift a complete sheet (which is normally installed in large sizes) you have to cut out a section and check for installed pipework and wiring before attempting any work. You should always try to cut chipboard flooring in positions where the sheet can be securely replaced on the joists.

Once the flooring has been lifted the next stage is to notch or drill the joists. The building regulations set out requirements for drilling or notching which must be followed at all times. An example is as follows:

> **Assuming the depth of the joist is 200 mm and the overall span of the joist is 2500 mm long**
>
> ■ The depth of the notch on the joist must be no greater than $1/8$ of its depth ($1/8$ of 200 mm is 25 mm)
> ■ The minimum distance away from the wall is 7 times the 'span' (2500 mm) divided by 100 mm.
> 7 x 2500 / 100 = 175 mm. This gives a minimum distance from the wall of 175 mm to start our notch.
> ■ The final distance required is the maximum area in which the notch can be made. This is worked out by dividing the span by 4 (2500 mm / 4 = 625 mm)

So with this knowledge we know we can safely make a notch 25 mm deep which must be a minimum distance of 175 mm and a maximum distance of 625 mm from the wall.

Notches are generally made with a hand or floorboard saw, hammer and sharp wood chisel. The notch should be large enough to allow the pipe to expand and contract especially when running hot water supply pipework. A section of pipe wrap can be used to help cushion the pipe onto the joist and a metal plate fixed over the pipe helps protect it when replacing the floorboards.

BENDING COPPER TUBE

To aid the clarity of view, all the bending techniques photographs featured here show the hand bending machine from above. You can fit the bender into a portable workbench or vice, but generally when you bend copper tube it's done with the machine simply held, or supported on the floor. These are all the basic bending techniques that you will require for most domestic installations.

1 When bending copper tube to a right angle, to fit around a corner for instance, mark the pipe at the required distance from the fitting into the corner or obstacle and place the measured end into the hook-side or tube stop side of the bender, ensuring a tight fit into the former (the former is the half round metal part of the bender in 15 mm and 22 mm sizes, in which the copper tube is placed and formed around).

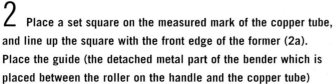

2 Place a set square on the measured mark of the copper tube, and line up the square with the front edge of the former (2a). Place the guide (the detached metal part of the bender which is placed between the roller on the handle and the copper tube)

between the roller and the copper tube and slowly pull the bender arm around until 90 degrees is reached (2b). Bend slightly over the 90 degree mark as when you release the arm the copper will bend back slightly.

3 To create two 90 degree bends from a fixed point, measure from the back of the tube (3a); this now becomes your fixed point. Mark the tube where you want the second bend to be. Place the copper into the bending machine and bend in the same way. Ensure that the first bend is located in the hook side of the bender, adjust the pipework so that the second bend will be in the correct direction in relation to the first bend (3b). Then place the guide between the roller and the copper tube and slowly pull the bender arm around until 90 degrees is reached (3c). Again bend slightly over the 90 degree mark to allow for the copper to spring back a little.

3a

3b

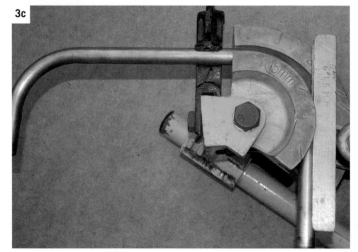

3c

TIP
When bending copper tube to fit into an awkward shape or corner, a template of the angle can be made by using a length of solder. This can be easily bent or shaped to create an accurate template to achieve the desired angle. This angle can then be used when bending the copper tube to achieve a very accurate and neat installation.

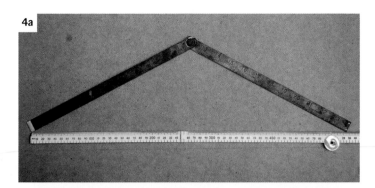

4a

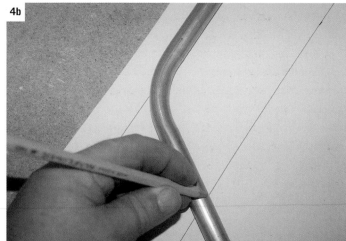

4b

4c

4d

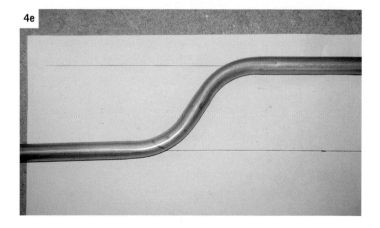

4e

4 To create an offset, use a 600 mm folding rule to determine a suitable angle to make the offset. First determine the height of the offset required, then subtract that figure from 600 mm. In the example shown here we have 75 mm – 600 mm = 525 mm. The folding rule can then be opened and set to this measurement (4a). This is now the first angle template. Place the tube into the bender and after fitting the guide, bend to the angle of the rule. Draw two parallel lines onto a piece of paper at the required distance of 75 mm and place the angled tube onto one of the lines, ensuring the tube is placed centrally. Mark the tube where it crosses the other line (4b). Place the tube back into the bender with the first angle fitted towards the hook, and the pencilled mark at a tangent to the former (4c). Bend the tube so it is parallel to the first angle (4d) or reuse the angle rule template. Place on the paper template to confirm the correct angle (4e).

5 When creating an offset to pass an obstruction the same methods are used but the distance from the fixed point to the first bend is needed. This is obtained by placing a temporary mark on the tube from the fixed point to the obstruction (5a), the height of the obstruction (offset), plus two tube diameters are added together and marked on the tube from the first temporary mark made. If the offset required was 50 mm and the tube used was 15 mm, then the total length measured back would be 80 mm (5b).

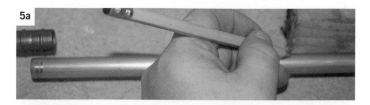

6 Once you have obtained this mark place the tube into the former ensuring the fixed point end is towards the hook. Using the folding rule at the required angle place it onto the bender in line with both tube and former, and place the pencil mark of the tube in the middle of the triangular shape made between rule and former. Then use the same techniques as for the first offset shown.

7 To achieve a full passover, create the first central bend by measuring the height of the obstacle you are passing over. Draw two parallel lines on a piece of paper to this distance. If making a passover from a fixed point then, measure from the fixed point to the centre of the obstruction and add a quarter of the obstruction diameter to this measurement (7a). In this case the obstruction is a 50 mm piece of pipe, so to gain an angle for the first bend, triple this measurement and then subtract it from 600 mm (the length of the folding rule). This gives us a length of 450 mm. Then, using the folding rule, place one end on the end of the tape and the other on 600 mm as used for the offset to gain the first angle (7b).

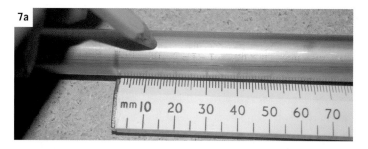

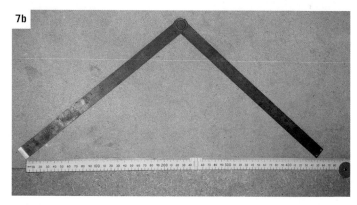

8 Place the tube into the former, with the fixed end towards the hook, and using the folding rule at the required angle place it onto the bender in line with both tube and former, and place the pencil mark of the tube in the middle of the triangular shape made between rule and former (8a). Pull the handle around until it is in line with the folding rule angle (8b).

8a

8b

9 The tube is now bent to this angle. Remove the tube and place it on to the paper template with the top line visible under the middle of the bend. Mark across the tube with a pencil as it crosses the bottom line (9a), turn the tube over and mark it again in the same way. Don't worry if the lines don't match perfectly. Replace the tube into the former with the first bend into the hook and the pencil line lying at a tangent across the former (9b). Fit the guide between the roller and former. Using either the folding rule unfolded or a straight edge, place one end onto the back pencil mark close to the hook (9c), and carefully pull the arm of the bender around until the tube is running centrally along the straight edge (9d).

9a

9b

9c

9d

10 Remove the tube and replace with the last bend close to the hook, again with the other pencil mark forming a tangent against the former (10a). With the guide in place and using the straight edge again, this time centrally placed along the tube. Pull the handle until the two lengths of pipe are in line (10b).

10a

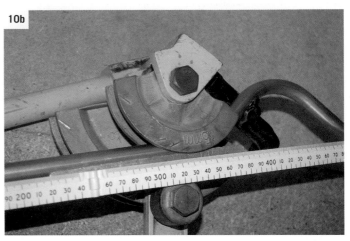

10b

11

11 To achieve a partial passover (a pipe that may pass over another copper tube and inserted into a tee or elbow), measure the distance required for the partial passover and then using similar methods as described earlier, subtract this measurement from 600 mm. In this particular case the height of the passover is 35 mm, so this would give a measurement of 565 mm. Place one end of the folded rule on the end of the tape and the other on 565 mm to create the first angle.

12 Place the tube into the former, and ensuring the guide is fitted correctly pull to this angle (12a). Double the first measurement taken to 70 mm and using the same procedure with the folding rule create another angle template. Draw two parallel lines on a piece of paper to the 35 mm passover distance. Place the angled tube onto this paper template and mark across the tube as it dissects the bottom line (12b).

12a

12b

13 Reposition the tube back into the former with the first bend towards the hook, ensuring the guide is in place and the pencil mark is lying at a tangent across the former (13a). Using the folding rule template, pull the handle around until the tube has formed this angle (13b). This end of the tube may require some trimming before it can be inserted into the fitting. If all the pipework is clipped the same distance from the wall, and then this can be achieved by placing a straight edge along the pipe and then cutting at this point (13c).

13b

13a

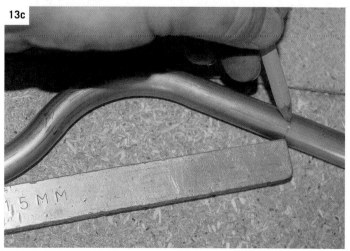

13c

TYPES OF SYSTEM

Before any isolation of pipework or any new installation or repair can be carried out, it is very important to know and understand the system that you are working on. These water systems provide a general outline of the different systems available and some time should be given to familiarise yourself with your own or your customer's systems.

COLD WATER SYSTEMS – DIRECT SYSTEMS

Beginning with the cold water systems of which there are two, the first is a direct system which is the most simple and is often found in many properties where either combination or pressurised hot water cisterns are used.

The direct system is when the cold water main serves all the cold water taps directly. If any CWSC and feed and expansion (F & E) tanks are present in the loft area, to supply a feed to the hot water and heating, then the float operated valves are also fed by this main supply (rising main). Unequal pressures are often a problem in bath shower mixers with the hot and cold being totally different pressures, resulting in either a very cold high pressure or extremely hot low pressure shower. In a large number of properties today either combination or pressurised hot water/heating systems are used, eliminating the need for storage cisterns in the loft area.

The main advantage of a direct system is that it is generally cheaper to install because of the reduced pipework. If a CWSC is required for the hot water cylinder, then it needs to be a minimum of 110 litres. Drinking water is available at all cold water taps, and with less pipework in the loft area there are fewer chances of pipework freezing.

The disadvantages are many with the high demand at peak times when the demand on the water supply is at its greatest which can result in low water pressure affecting the performance of showers and water heaters. There is also extra wear and tear of the taps and valves due to the high water pressure. The pipework can be noisy due to the high water pressure, and if the main cold water is shut off, then no stored water is available.

COLD WATER SYSTEMS – INDIRECT SYSTEMS

With the indirect system the rising main only connects to the kitchen tap, and then continues rising to feed the CWSC and possibly the F & E if present. There is then a down service from this cistern which feeds the bathroom and any other cold water tap (e.g. cloakroom or additional WC). If the cistern is large enough (minimum 225 litres), then possibly a hot water cylinder cold feed would also be connected.

The main advantage of the indirect system is that if the mains are turned off, there will be a stored supply of water available. There is also a lower demand at peak times and reduced noise in system pipework along with reduced wear and tear on taps and valves.

The disadvantage is the higher cost of installation along with an increased risk of freezing pipework and larger CWSC.

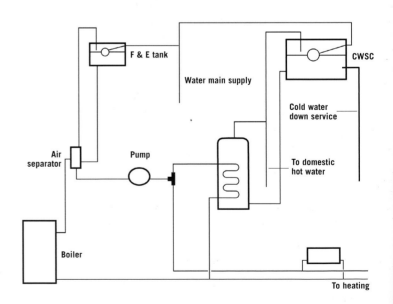

Indirect cold and hot water system

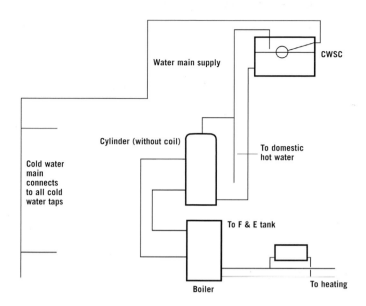

Water main supply

CWSC

Cylinder (without coil)

To domestic
hot water

Cold water
main
connects
to all cold
water taps

To F & E tank

To heating

Boiler

Direct cold and hot water system (vented)

HOT WATER SYSTEMS – DIRECT SYSTEMS

Direct hot water systems are generally heated by either instantaneous multi and single point water heaters or hot water cylinders which are fed from the CWSC. If either a gas or electric instantaneous heater is used, then an isolation valve (stopcock or quarter turn ball type) should be installed close to the heater, to allow for maintenance. Direct hot water systems (vented) hot water cylinders are also used in conjunction with boilers, with the water heated from the boiler rising up the primary flow pipe to the cylinder, due to convention currents in the hot water. This system is generally known as a gravity circulation system. This then heats the water directly in the cylinder and as the cooler, heavier water falls it returns to the boiler through the primary return pipework. The process is repeated until the required water temperature is obtained. This temperature should never exceed 60–65° C (140–149° F).

This type of cylinder can also be heated by an immersion heater. The primary flow and return pipes for these systems is generally 28 mm especially if it's a continuous burning appliance. Twenty two mm pipework is only used if the distance between boiler and cylinder is very close. The vent pipework must be 22 mm and the vent route forming the primary pipework should not be valved. Corrosion inhibitor must not be used in the CWSC.

HOT WATER SYSTEMS – INDIRECT SYSTEMS

The most common types of indirect hot water (vented) system are generally heated by means of a heat exchanger coil inside the cylinder. The water inside this coil is heated from a boiler through primary flow and return pipework and is kept completely separate from the water inside the cylinder.

An additional F & E cistern is used in this system and because the waters are kept separate, an inhibitor can be used in the F & E cistern. The other type of hot water heating is the single feed indirect system. This is a self-venting cylinder and does not require a separate F & E. The water between boiler and cylinder is kept separate by an air bubble which is located inside a heat exchanger in the cylinder. Inhibitor must again never be used in this type of system. This system should only be installed to the manufacturer's instructions as failure of the air bubble means the waters will mix, contaminating the hot water. If the air bubble has failed then a typical sign is brown rusty water appearing from the hot water taps. This type of hot water heating is not widely used anymore, but you may still come across it.

There are a number of different ways of heating water, these can be natural or liquid petroleum gas (LPG), solid fuel and oil boilers. Gas can also be used on water circulators, instantaneous water heaters and storage heaters. Oil and solid fuel can be used with combined cooker and boiler appliances. Electricity is the other main energy source for heating water by using either immersion heaters, instantaneous water heaters or storage heaters. Solar panels or collectors are an environmentally friendly method of heating water and are usually used to help supplement other hot water systems.

TESTING PROCEDURES

Testing for available pressure and flow rate is very important, and determines at what pressure pipework should be tested and what appliances or fittings can be installed. Instantaneous water heaters and combination boilers require certain pressure and flow rates to enable them to work efficiently. Many modern taps now require a high pressure for them to work properly, and without proper assessment of water pressure and flow rates expensive errors can be made.

WATER PRESSURE AND FLOW RATES

To test the incoming water pressure connect a small pressure gauge to the kitchen tap or other mains supplied by either a screw connector or rubber type push fit connector. With the water turned fully on read the pressure gauge, which is generally marked in bar pressure **(1)**. This test should be carried out at both peak and off peak times to ensure minimum and maximum readings are obtained.

Testing for flow rates is simply done, using a weir gauge placed under a full flowing tap **(2)**. A time is taken of the amount of water delivered through the gauge. This is recorded as litres per minute. Flow rate tests should also be taken at both peak and off peak times to ensure minimum and maximum readings are obtained.

PRESSURE TESTING

Once you have completed the pipework installation it is advisable to pressure test it prior to filling with water. All these testing procedures are carried out using a hydraulic pressure tester, which can be hired from most hire centres. These are oblong bucket-type containers that have a pressure gauge, hose and pipe connector fitted to one end. The water is then pressurised into the pipework by means of a lever handle and piston which is positioned close to the gauge. When the handle is pumped it forces the water along the hose and into the pipework. It has a non-return device fitted to prevent the water re-entering the pressure tester.

This type of testing procedure is generally used on the first fix pipework. This is the pipework installed under floors or ductwork and is the supply pipework to all appliances. This pipework is best tested before replacing flooring or ductwork, enabling all joints to be examined.

Ensure all first fix pipework is securely fitted and that all open ended pipework has either a valve or cap securely connected before testing begins.

The test procedure which should be used is set out in the British standards of 6700. These state that two separate procedures are used for either rigid (e.g. copper) or plastic pipework. The testing of rigid pipework is that once the pipework has been completely filled, the water temperature should be allowed to stabilise for thirty minutes. The pipework should then be pressurised to one

2

and a half times its working pressure. After a period of one hour, if no leaks are present, then the pipework is satisfactory. If any leaks are found, then the same procedure is carried out again after a repair has been effected.

The testing of plastic pipework is more complicated and there are two procedures that can be used. Procedure A is to apply a test pressure by pumping water for 30 minutes and visually checking for any leaks. Once this has been done, reduce the pressure from the system to 0.33 times the maximum working pressure. Examine and check the pipework and if after 90 minutes the pressure remains at or above 0.33 times the working pressure the system is regarded as satisfactory.

Procedure B is to apply a test pressure of one and a half times the working pressure and to continue pumping water for 30 minutes taking note of the pressure and visually inspecting the pipework for leaks after this period. If after a further 30 minutes the pressure drop is no less than 0.6 bar, then the pipework is considered to have no obvious leaks.

Visually check the pipework and gauge for an additional 120 minutes. If after this time the pipework has dropped to no more than 0.2 bar then the system can be regarded as satisfactory.

The soundness and performance testing procedures for soil and vent is covered in Waste systems (see pages 70–71).

TRAP SEAL LOSS

There are a number of ways that trap seal loss can occur. These include induced and self siphonage, compression, momentum, capillary attraction, wavering out and evaporation.

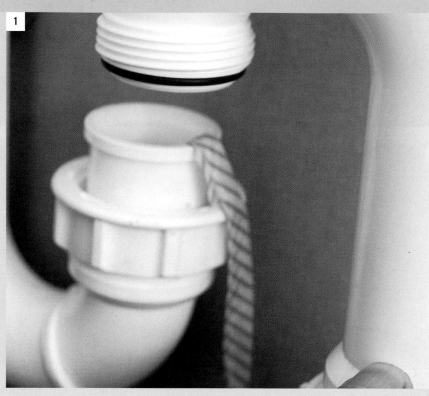

Induced and self siphonage, along with compression, would be the cause of seal loss after the performance test. Induced siphonage happens when water is discharged from another appliance connected to the same waste pipe. Negative pressure is created inside the pipe as it runs full bore. This problem can be overcome by either adding an additional vent pipe to the system or, for the easier option, by fitting a resealing trap.

Self siphonage often occurs particularly in basins. As water discharges from the basin a plug is again created causing a negative pressure and a partial vacuum in the pipe which siphons water out of the trap as it falls. This again can be overcome by the use of a resealing or anti-vac trap.

Compression can occur when an appliance, usually a WC is flushed on a first floor, and as the water falls it compresses at the base of the stack causing a back pressure which has enough pressure to force the water out of the trap at a lower level. This generally occurs when the stack has been installed incorrectly without the use of a long radius bend. All these problems are usually caused by poor installation practice.

There is usually a simple remedy for other types of trap seal loss. With momentum the seal could be lost by a large amount of water, perhaps from emptying a bucket into either a WC or sink, thus removing the trap by the momentum or force of the water. Re-sealing trap can be used to help prevent this in the basin.

Wavering out is the term given to wind blowing across the top of an open vent pipe, causing a wave in the WC which in turn causes the water to be lost over the weir of the trap. This can be remedied by the addition of a bend or cowling fitted to the top of the vent pipe.

Capillary attraction (1) can be the cause of trap seal loss if a piece of string or thin material has become stuck in the trap, acting as a wick and removing the water seal in the trap. This can often happen in washing machine traps. The only solution to this problem is to ensure that any loose material is cleaned away from the plug waste area and that no loose material is placed inside the washing machine.

Evaporation is the final trap seal loss which will only occur if the appliance remains unused and through natural evaporation the seal is lost. The only solution here is to let a small amount of water run into the waste system. An example would be to let water run into an appliance which is hardly ever used, once every month to help maintain the trap seal.

TRAPS

Traps prevent smells entering the premises. As water is discharged into the waste it passes through the trap to the sewer system, retaining a specific amount of water to prevent any smells entering back through the trap.

'S' traps are generally used when the waste pipe has to go straight to the floor for a low level connection and the 'P' trap can discharge directly behind or with a swivel elbow be converted into an 'S' trap.

Bottle traps are treated as 'P' traps and can also be converted to an 'S' trap if required. 'S' traps can be a problem with trap seal sometimes being lost through the sudden drop (self induced siphonage) so an anti-vac bottle type trap is a good alternative. And although bottle traps are good for basins they should be avoided on kitchen sinks because of food and other materials getting blocked in the bottom of the trap.

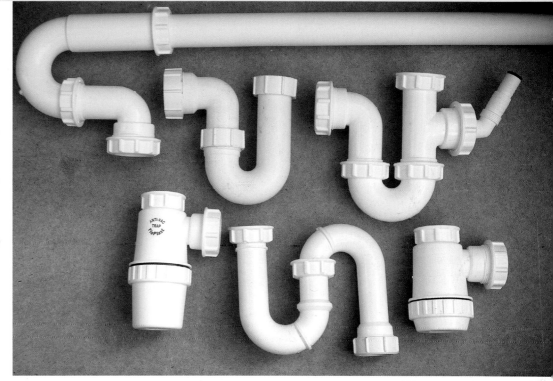

Top row, from left to right: washing machine and standpipe; 'P' trap; 'P' trap with washing machine connector. Bottom row, from left to right: anti vac trap; 'S' trap; bottle trap.

Bath traps are designed to be low level to fit in tight spaces below the bath and they usually have a trap of only 38 mm.

A washing machine stand pipe is very simply a 'P' trap with an extended length of waste pipe to take the corrugated waste pipe of a washing machine or dishwasher. The alternative to this connection is to install a 'P' trap with a washing machine spigot.

Running traps are generally used in domestic premises where 'S' or 'P' traps can't be used and are often used in waste pipework for washing machines or dishwashers.

The table below shows the size of waste fittings and possible traps and trap size:

Appliance	Waste fittings (in)	Trap type	Trap size (mm)
Bath	1½	Bath trap	42
Basin	1¼	S, P, bottle, straight through	32
Bidet	1¼	S, P, bottle, straight through	32
Sink	1½	S, P	42
Washing machine/dishwasher	1½	W/M stand pipe P with spigot	42

CLIPS, BRACKETS, SCREWS AND FIXINGS

When installing appliances and pipework it is very important to know the correct distance between clips and the correct methods of fixing radiators and sanitary ware.

Supporting pipework is extremely important in the installation of all plumbing projects. Clips should always be fixed in place prior to pipework installation, unless circumstances do not allow this. In most domestic installations of pipework a number of different types of plastic clip that are either open or hinged with a wrap-over top are used (1). These can be purchased in either double clip style or have the ability to be clipped to each other to form a bank of clips for group runs of pipework.

Another type of plastic clip available is very similar to electrical clips. These clips are hammered into the wall by means of a masonry nail (2). These are generally used for supporting pipework which has to be very close to the wall and is often used with plastic or copper pipework.

The copper saddle clip (3) is designed for exactly the same purpose, but generally only for copper pipe and is often found in old properties. It generally only supports pipework which is close to the wall but can be used for large pipes.

The other type of clip that can be used especially when the pipe has to be insulated with a thick insulation to prevent freezing or heat retention is the ring and back plate. This consists of a brass ring with a screw connection to each side, and can be extended further away from the wall by means of a threaded rod and female back plate (4). This is a very versatile clip which can be combined to each other by means of additional threaded rods, enabling multiple pipework to be installed in restricted areas.

A similar type of ring and back plate is also used for low carbon steel pipework along with skirting clips that enable the heavy pipework to be laid onto them before the top part of the clip is screwed in place.

The clips for supporting plastic waste pipework are generally of the saddle type variety and are installed during the installation of the pipework (5).

When clipping hot and cold water services the hot water should always be fixed above the cold, to prevent heat transference. Water supplies should also be run at least 25 mm from all gas supplies and 150 mm from all electrical services.

The recommended clipping distances are for all pipework is shown in the table left.

Pipe size		Copper		LCS		Plastic pipe	
		Horizontal	Vertical	Horizontal	Vertical	Horizontal	Vertical
mm	in	(m)	(m)	(m)	(m)	(m)	(m)
15	½	1.2	1.8	1.8	2.4	0.6	1.2
22	¾	1.8	2.4	2.4	3.0	0.7	1.4
28	1	1.8	2.4	2.4	3.0	0.8	1.5
35	1¼	2.4	3.0	2.7	3.0	0.8	1.7
42	1½	2.4	3.0	3.0	3.6	0.9	1.8
54	2	2.7	3.0	3.0	3.6	1.0	2.1

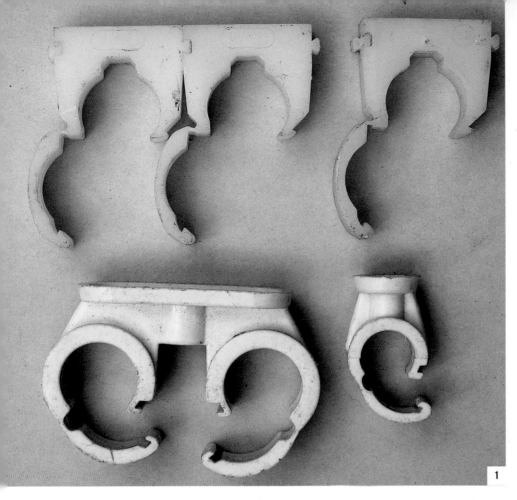

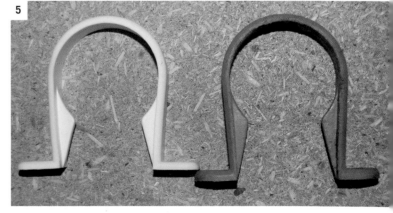

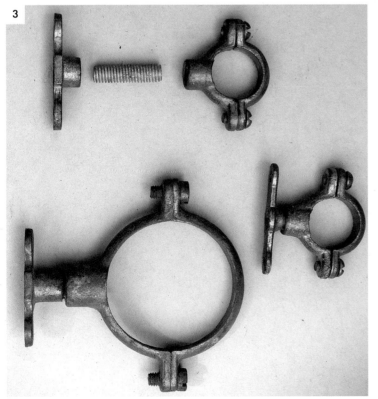

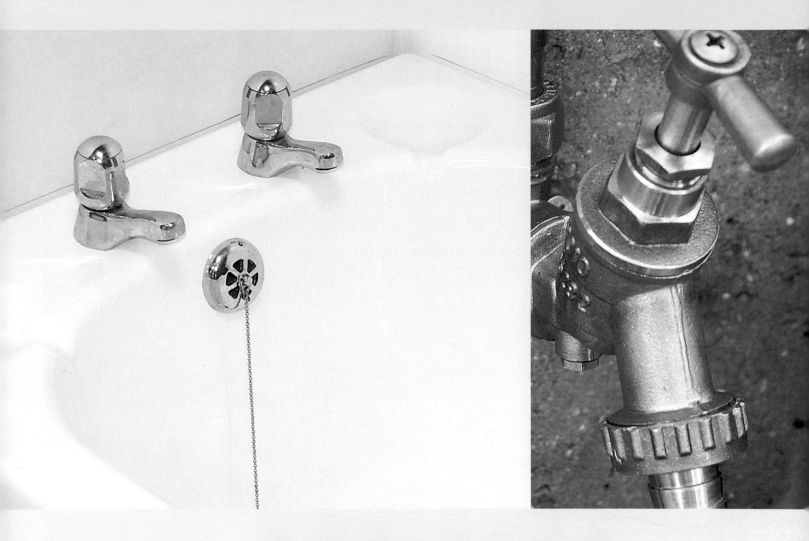

PROJECTS

INSTALLING A BASIN

This is a typical installation of a basin and pedestal comprising simple pillar taps and a slotted waste assembly. An alternative wall-hung basin, which is often used in additional cloakroom areas, is also shown along with a typical 'mono bloc' tap and pop-up waste assembly. Copper is used in both basin installations, but all supply pipework can be made of plastic.

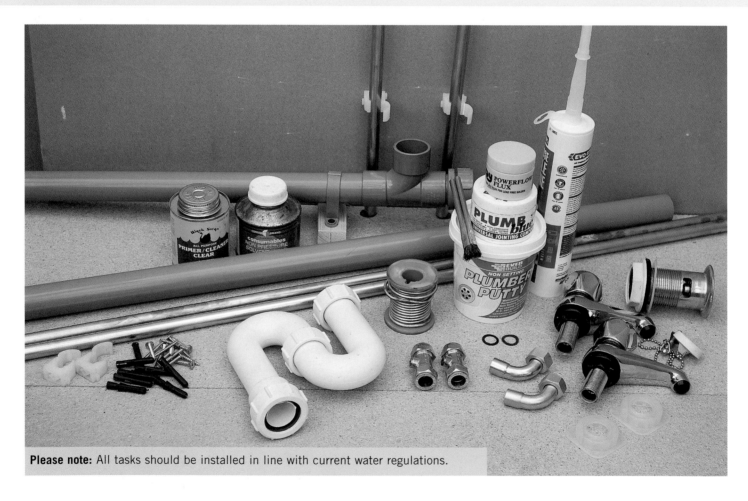

Please note: All tasks should be installed in line with current water regulations.

For a typical basin and pedestal, the following materials are required: slotted waste, plug and chain, basin taps, tap connectors and fibre washers, top hats, 15 mm copper tube, 15 mm clips, 15 mm bends, isolation valves of the quarter turn type, silicone sealant, screws and plastic wall plugs, lead-free solder, flux, 32 mm 'S' trap, 32 mm plastic waste pipe, solvent cleaner and glue, potable/wholesome jointing compound.

For the wall-hung basin, the following materials are required: slotted waste, plug and chain, basin taps, basin brackets, tap connectors and fibre washers, top hats, 15 mm copper tube, 15 mm clips, 15 mm bends, isolation valves of the quarter turn type, plumber's putty screws and plastic wall plugs, lead-free solder, flux, 32 mm bottle trap (or adjustable bottle trap), potable jointing compound.

■ The basic preparation for this job consists of knowing the location of the hot and cold service valves and their working condition, and whether it is possible to isolate either the bathroom basin or the whole of the hot and cold water supply.

■ For all the projects in this book it is established that the hot and cold water supply has been completely shut down and no independent isolation valves are fitted. Within the installations of the basins new independent isolation valves of the quarter turn type are installed.

■ A selection of plastic and material type dust covers should be used with the plastic completely covering the floor area along with the material sheet on top.

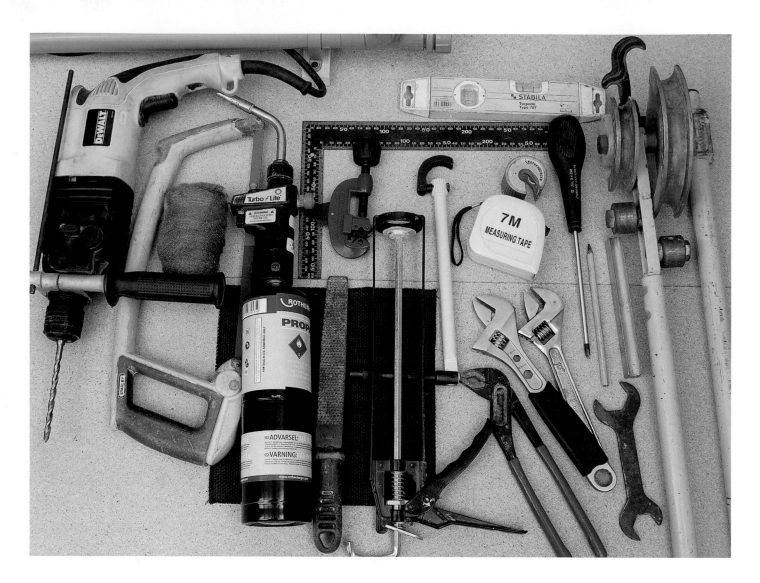

The tools required for either job consist of: two adjustable spanners, 15 mm pipe slice (or tube cutter), pipe bender and 15 mm guide, long cross head screwdriver, gas torch, heat mat, wire wool, pump pliers, basin spanner, flat spanner, rasp or file, tape measure and pencil, hacksaw, drill, boat level.

BASIN AND PEDESTAL INSTALLATION

1 You need to ensure that if the basin has an integral overflow (1a) within its construction, then a slotted and not a solid waste assembly is used. If it is a twin-holed basin with taps, washers and back nuts, it should have a plug (rubber or chrome) and chain. Conversely, if it is a single hole, this is a mono block style tap, and should have a pop-up waste (1b).

> A slotted waste allows water to enter the trap from an internal overflow, which is manufactured in the basin. A solid waste is used when no internal overflow is present.

1a

1b

2 Place the basin on its back and apply clear silicon around the outlet on the waste (2a). An alternative to the clear silicon is a plumber's putty, which can be applied in a similar way as glazing putty (2b), either spread around the waste outlet on the basin or the slotted waste. The slotted waste is then passed through the basin outlet ensuring a firm contact between basin and waste.

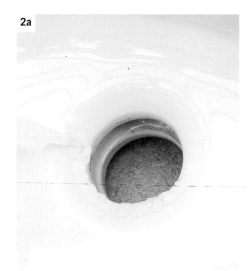

2a

2b

3 Turning the basin carefully over again apply either some more silicon sealant or putty around the waste and the top threads (3a), then using the washer and back nut make a firm connection, tightening with pump pliers (3b). Remove any excess sealant or plumber's putty around top threads and the waste with a pointed wooden rod (3c), then use a damp cloth to wipe off the sealant. If the basin does require a plug and chain, then the sealant or putty can be used to achieve a seal and any residue removed.

While the basin is still in this position, push the hot tap through the left tap hole, as the basin is facing you. A small bead of silicon or putty around the tap hole of the basin before the tap is fully pushed through helps seal the tap.

3a

3b

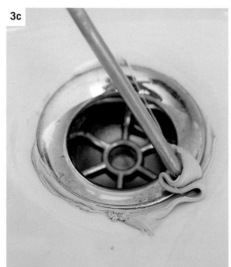
3c

4 Underneath the basin where the tap protrudes through the hole, place a top hat (4a) on to the threaded tail of the tap, then using a basin wrench, tighten the back nut until it is in a fixed firm position (4b). Repeat the same sequence with the cold tap. The basin is now connected with taps, waste and plug.

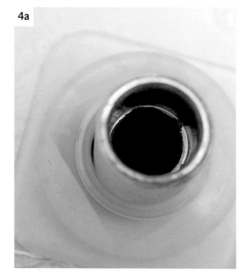
4a

4b

5 Here is a typical pipework installation (5a). As stated earlier, there are no independent valves for the basin, and although the water regulations don't ask for valves to be fitted to basins it is good practice to fit isolating valves to aid future maintenance and make the initial installation easier, with the installer being able to test each service and isolate quickly if a problem or leak occurs.

The waste pipe is terminated within the bathroom area, so a simple waste connection at low level will be made. The supply pipework, held in place by a clip on each pipe, emerges from the floor with the hot water supply on the left hand side; this is standard practice with all hot water taps installed on the left, whether on a basin, bath, shower or kitchen sink. Fix either chrome or brass isolating valves of the quarter turn ball type (5b). The types shown here are compression fittings, and are installed using methods described in Joining methods (see pages 31–36).

At this point and ensuring that the fittings are tight and that the valves are turned off and are facing to the outside of the pedestal edge (5c), the water supply can be turned back on and tested up to the isolation valves.

5a
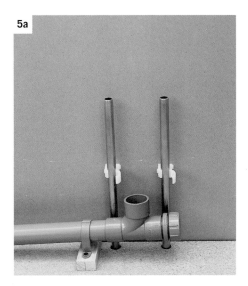

5b
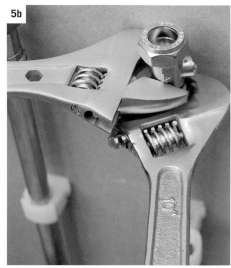

5c
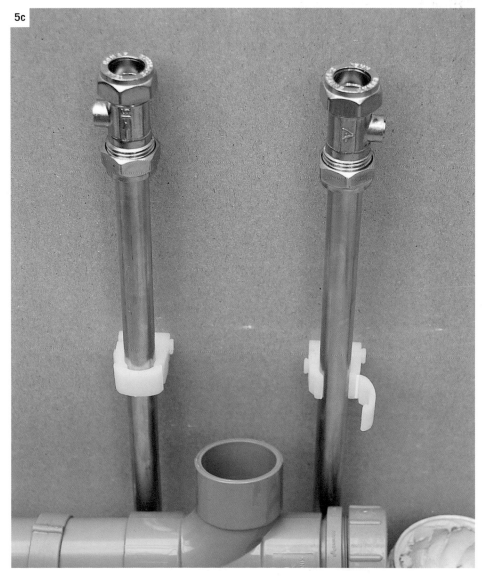

6 With the aid of the small boat level, mark and fix two further clips approximately 150 mm above to help support the pipework and to help prevent any pipe noise which may occur.

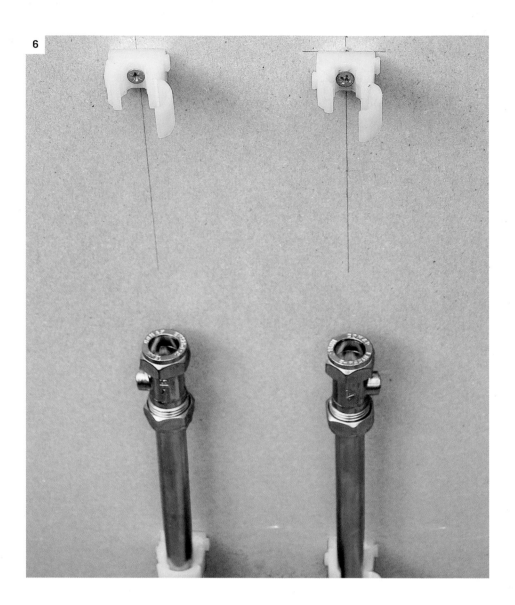

7 Place the pedestal in position (the basin pedestal is always a distance from the wall, and this changes with each different basin design), ensuring that it fully covers the supply and waste pipework (7a). Then, carefully place the basin on top of the pedestal and make sure that it is both level and fits flush to the wall. Use a boat level for this (7b).

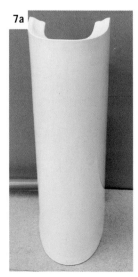

8 Using a pencil, mark the wall through the two holes in the bottom of the basin. These holes are situated under the basin at either side, then remove the basin to a safe place. Mark lightly around the base of the pedestal with a pencil before also removing it to a safe place. This will help you to place the pedestal back in the same position, after the drilling of the wall has been achieved.

9 If you are fixing to a solid wall, the holes drilled will have to be plugged. If fixing to a cavity wall, proprietary cavity fixings must be used, as described in Fixings (see pages 52–55). For most basins, the hole to help support the basin is drilled at a slight angle upwards (9a). If drilled level it will not line up with the basin holes correctly, so with this in mind much care should be taken when drilling these holes.

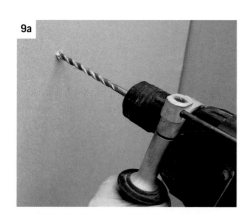

Reposition the pedestal making sure that it lines up with the pencil mark on the floor, then fix it securely to the floor with approximately a 32 mm (number 8 or 10) large-headed screw that has been wrapped in putty, to help prevent damage to the pedestal (9b). Before screwing into the floor care must be taken to prevent screwing into hidden pipework or electric wiring, which may be installed under the floor (use a pipe and electric cable detector for this purpose). When screwed securely in place, place the basin over it and securely fix it to the wall, again wrapping the screws in putty. With the use of a boat level, ensure that the basin is still level.

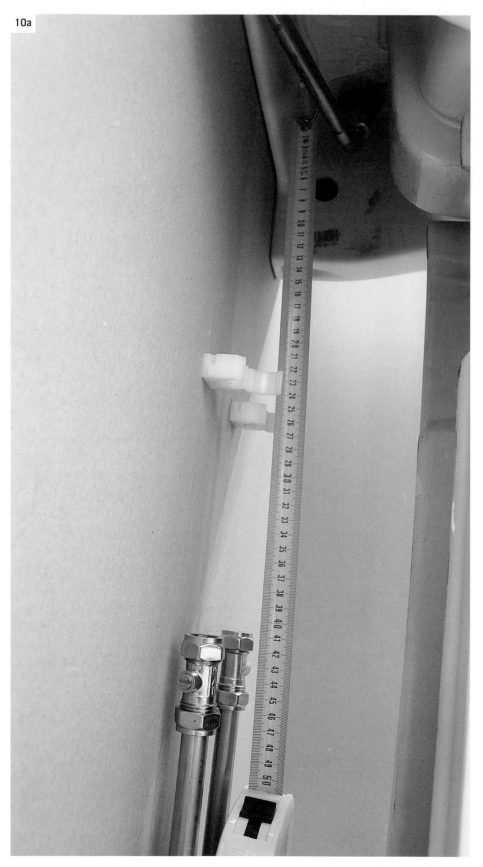

10a

10 Once the basin is securely fixed to the wall, fit two angled tap connectors to the bottoms of the taps and place a short piece of pipe in each. Measure from the internal side of the isolation valve up to these short pieces of pipe (10a) and using the method described in Bending copper tube (see pages 59–65) bend two pieces of pipe to 90 degrees (10b) at those measurements. Remove the short pieces of pipe from the tap connectors and replace with the angled pieces of copper. Trim the copper to fit, again ensuring that the pipe inserts fully into its fitting. An alternative to bending the copper tube is to use soldered push fit or compression bends.

10b

11 Remove the angled copper tube along with the tap connectors. Solder the tap connectors to the pipe (11a) as described in Joining methods (see pages 31–36), ensuring that the fitting and pipe are in a level position (11b)

Once the fittings have cooled down, the next process is to connect the tap to the isolating valve. One of the 'tricks of the trade' is to lightly apply potable jointing compound to the seating of the tap connector, before fitting the fibre washer (11c) and then applying a second light coating on top of the fibre washer. This stops the washer falling off when trying to connect to the taps, and helps create an extremely waterproof seal between the tap connector and the tap.

11a

11b

11c

12 Refit the pipes into the isolation valves but don't fully tighten the nuts; just gently hand tighten at this stage. Then fit the tap connector onto the tap ensuring that the fibre washer is still in place, then just using your hand and ensuring that the connector is not cross threading, tighten until the connector is on as far as it will go by hand. Then using a basin wrench or crow's foot, fully tighten the tap connector then the isolation valves. Repeat the procedure for the other side. The supply pipework is now complete.

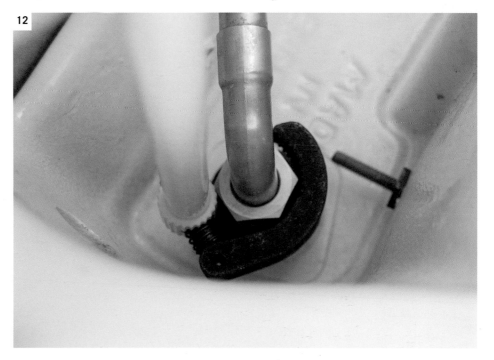

12

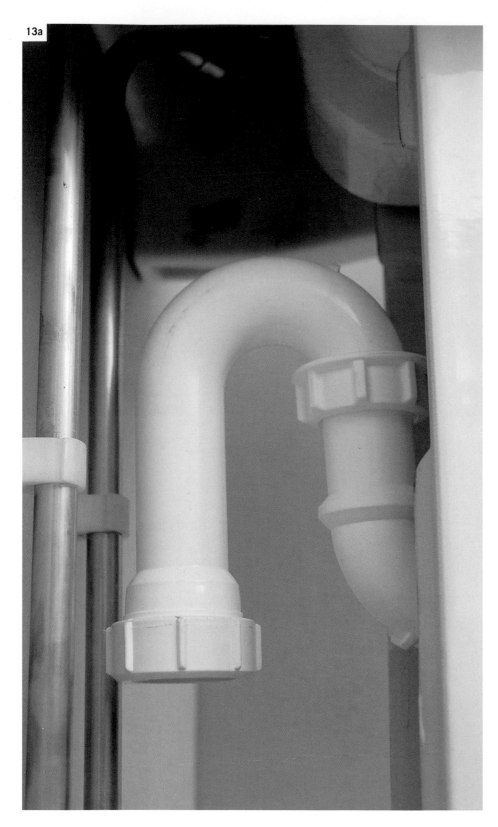

13 The last part of the process is to fit the 'S' trap onto the waste (13a), then measuring from the internal part of the trap to the internal part of the fitting (13b), a length of 32 mm plastic waste pipe can then be cut to size.

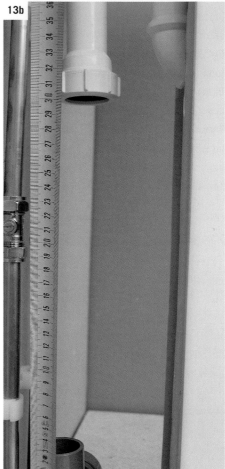

14 Remove half the 'S' trap leaving half still connected to the waste (14a), then connect the cut length of 32 mm plastic waste pipe into the trap, (14b), deburr the other end of the waste pipe (14c) and clean with wire wool and a proprietary cleaner. Apply solvent glue to the pipe end and tee fitting (14d and 14e) as described in Joining methods (see pages 31–36). Then fit the pipe into the tee connector (14f) and make the final connection by joining the 'S' trap back together (14g). Fix earthing connections to pipework as described in Earthing (see pages 20–21). Test both hot and cold water supplies individually, along with the new trap and waste system, let the water run until clear. The basin is now complete.

14a

14b

14c

14d

14e

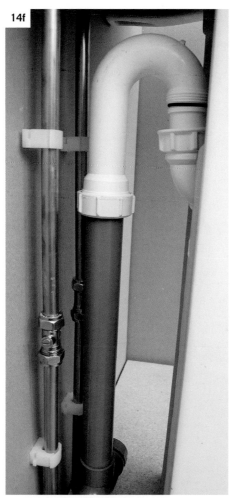

14f

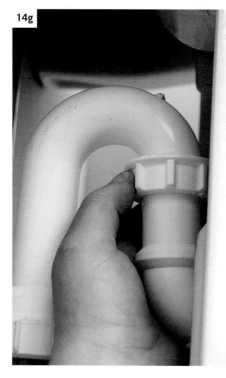

14g

WALL-HUNG BASIN INSTALLATION

1 For a wall-hung basin the typical height is approximately 800–850 mm from the floor level. If as in this installation the waste pipe is already in place, then a measurement must be gained to establish the height of the basin from the waste pipe. Alternatively a telescopic bottle trap (a trap that can be adjusted up or down to line up with the waste pipe) can be used.

2 Fitting brackets for this type of basin can be tricky. Hold the basin at a mark on the wall at 800 mm and as you are holding with one hand, use a pencil to mark on the wall the bottom of the basin with your other hand (2a). Place the basin on the floor and measure the distance required between the brackets (2b) and fix a bracket at that point on the wall. Measure the determined distance and with the use of a spirit level, fit the other bracket.

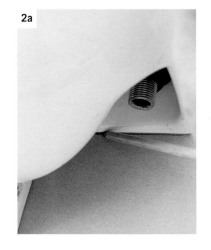

3 For the wall-hung basin we are going to use an adjustable 'bottle' trap. Using similar methods for fitting the waste and taps as in the basin and pedestal project, once the brackets are fitted on the wall the basin can then simply be hung. Before proceeding any further, ensure that the basin is level, that the isolated valves are on the hot and cold supply pipes, and the clips are in place to secure the new pipework.

4 Before fitting the supply pipework the trap assembly can be connected. The waste pipe in this installation terminates through the wall, so a measurement is taken from the extruding waste pipe into the full depth of the trap, (4a) and the pipe is marked and cut in situ. After burring and cleaning the waste pipe as described for the basin and pedestal, connect the trap onto the plastic waste pipe and then onto the slotted waste of the basin (4b).

5 Now measure the distance between the hot pipe and the straight tap connector. Using a short piece of copper tube inserted into the isolation valve and another piece inserted into the tap connector, a measurement can be obtained between the two (5a) and used to create an offset in the copper pipe (5b). This can be achieved by following the bending procedures as described in Bending copper tube (see pages 59–65).

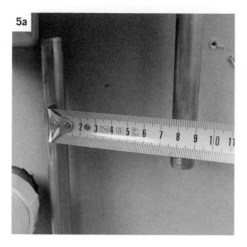

5a

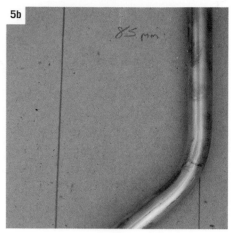

5b

6 Place the offset pipe into the isolation valve, then mark (6a) and trim the pipe as required, again making sure the pipework will fit fully into both the straight tap connector and the isolation valve (6b).

6a

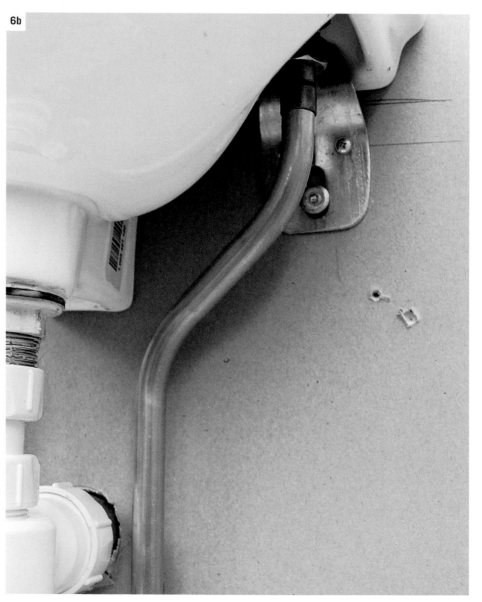

6b

7 Remove the pipework and the straight tap connector, and solder together as with the basin and pedestal installation (see Step 10a, page 86). When cooled, use the same fibre washer and potable jointing compound techniques to refit the pipe. This is refitted into the isolation valves first but don't fully tighten the nuts, just gently hand tighten at this stage. Then fit the tap connector onto the tap ensuring that the fibre washer is still in place. Once secure, use your hand to tighten and take care not to cross thread the connection.

As in the basin and pedestal, tighten until the connector is on as far as it will go by hand, then using a basin wrench or crow's foot (see Step 12, page 87) fully tighten the tap connectors and then the isolation valves. Repeat the procedure for the other side. Fix earth connections to pipe work as described in Earthing (see pages 20–21). The basin is now complete.

Don't forget to test both hot and cold water supplies individually and let the water run clear. Test the new trap and waste system with the plug in and the plug out.

INSTALLING A BATH

This is a basic installation procedure for a typical plastic bath with simple single bath taps along with a combined overflow and waste assembly. The same installation procedures are followed for steel baths, with the exception of the supporting legs which are either stuck to the bath or attached by a threaded rod and wing nuts connected to preformed fixings but follow manufacturer's instructions for this. Steel baths must also be earthed.

Here, a silicone sealant has been used for the waste connections, but plumber's putty can be used instead. Copper is used in this installation but all supply pipework can be installed using plastic pipework and fittings.

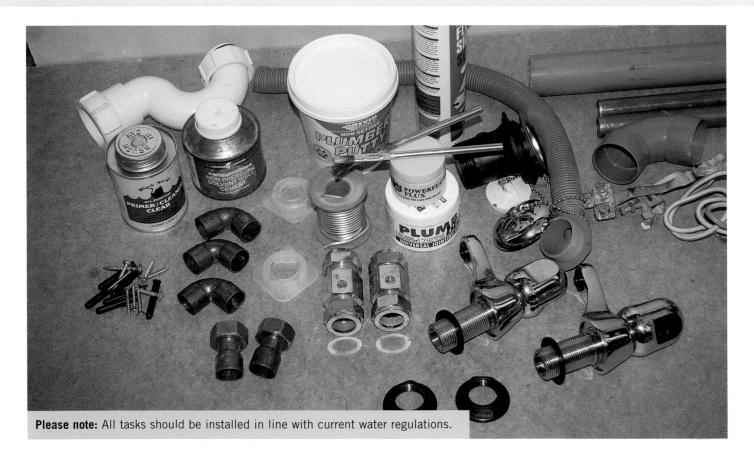

Please note: All tasks should be installed in line with current water regulations.

For a typical plastic or steel bath, the following materials are required: combined waste and overflow assembly, plug and chain, bath taps (singles, deck mixer or mono bloc), tap connectors and fibre washers, top hats (only if single taps are used), 22 mm copper tube, 22 mm bends, isolation valves, non-return valves (if connected to a direct cold water system), silicone sealant (or plumber's putty), screws and plastic wall plugs, lead-free solder, flux, 40 mm bath trap, 40 mm plastic waste pipe, 40 mm plastic bend, solvent cleaner and glue, potable jointing compound, earth straps and earthing wire.

The following tools are required: two adjustable spanners, 22 mm pipe slice (or tube cutter), pipe bender (and 22 mm guide), long cross head screwdriver, gas torch, heat mat, wire wool, pump pliers, basin spanner (basin spanners are used on all tap assemblies except mono bloc type), flat spanner, rasp or file, tape measure and pencil, hacksaw, drill and level.

■ The basic preparation before attempting this job, or any project in this book, consists of knowing the location of the hot and cold service valves and their working condition, and whether it is possible to isolate either the bath or the whole of the hot and cold water supply.

■ For the projects in this book, it is established that the hot and cold water supply has been completely shut down, and within the installation of the bath new independent isolation valves of the quarter turn type are installed.

A selection of plastic and material type dust covers should be used with the plastic completely covering the floor area along with the material sheet on top.

1 Before installation of the plastic bath place the bath upside down on either soft dust sheets or an equally suitable material to avoid damaging the bath. Remove all the packaging and place the supporting legs across the bottom base board at determined measurements from the waste outlet and board end (refer to installation instruction, which can differ from bath to bath). Before securing the supporting legs to the base board make sure you are using the correct screws supplied with the bath and NOT the transit screws that supported the legs in packaging. If the wrong size screws are used there is a great possibility they will penetrate the

bottom of the bath. This is a very expensive mistake to make so great care should always be taken at this early stage.

2 Once both supporting brackets are fixed to the base of the bath and the supporting framework (2a), fit the centre leg (2b) between the two, taking care again to use the correct screws supplied with the bath.

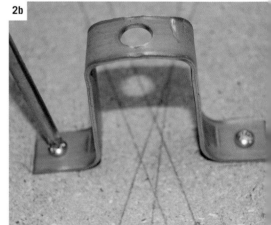

3 Fit the adjustable feet into the supporting legs and adjust the level of the feet by measuring from the top of the bath to the bottom of the adjustable feet. Baths are generally installed at a height of between 500 and 600 mm. This measurement is usually determined by the height of the bath panel or the height of the waste outlet.

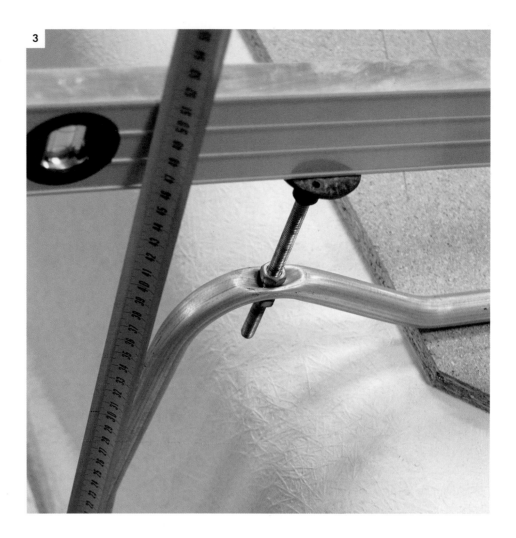

3

4 Once the legs and feet have been firmly secured turn the bath over onto its feet. Apply silicon sealant or plumber's putty around the inlet waste hole (4a). The waste is then pushed firmly against the sealant. Apply more silicon sealant on the underside of the bath where the waste outlet comes through. The washer is then placed onto the waste along with the overflow connector and back nut and firmly secured. Wipe off any residue of sealant with a damp cloth before proceeding with the overflow connection (4b).

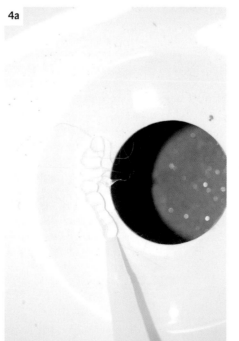

4a

4b

5 Apply a small amount of silicone or plumber's putty around the overflow on the external side of the bath, and push the overflow connection through on to the sealant and into the inside of the bath (5a). Apply another small amount of sealant on to the front of the overflow connection, and tighten on to the threaded overflow connector (5b). Remove any residue of sealant from the overflow, but don't fit the corrugated overflow pipe connection at this stage, because during the soldering of the supply pipework this pipe can be easily burnt and melted. The overflow connection must be sealed very securely as once the bath is installed achieving a repair can be very difficult.

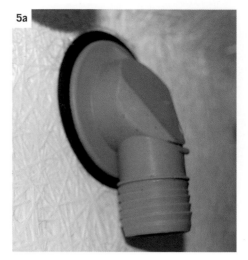
5a

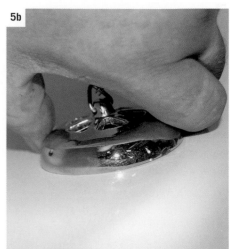
5b

6 While the bath is still in this position push the taps through the tap holes, ensuring that the hot tap is placed on the left hand side as the bath is facing you. A small bead of silicon or putty applied around the top hole of the bath before the tap is pushed through helps seal it in place. Place a top hat onto the threaded tail of the tap, where it protrudes underneath the bath (6a), then, using the back nut, tighten until the tap is in a fixed position (6b). Repeat this sequence with the other tap. Connect the plug and chain to the overflow outlet (6c) and connect fixing brackets to the wooden supports of the plastic bath (6d).

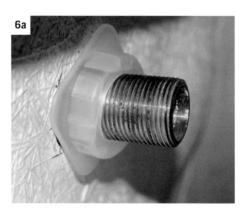
6a

6b

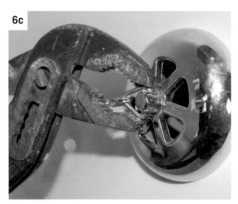
6c

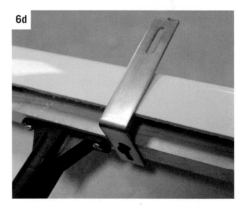
6d

The bath is now complete with legs, taps, waste, overflow and plug. The only additional fittings that may be required on some baths are the handles. The installation of these depends on the design, but care must be taken when fitting them to ensure they are fixed securely, as tightening a handle which is close to the wall after installation can be very difficult!

7 Not all baths require handles. Here, however, the handle is fitted by means of a threaded rod screwed into each end of the handle and fixed by securing a grub screw (7a). The metal rods are passed through the holes in the bath with a plastic washer fitted to each end of the handle. A nut secured on the underside tightly holds each end of the handle in place (7b). Repeat the same procedure for the other handle.

7a

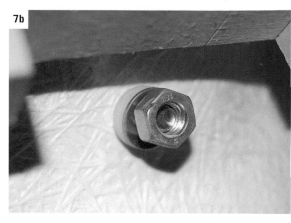

7b

There are no independent valves for the bath, and although the water regulations don't ask for valves to be fitted to the bath it is good practice to fit isolating valves to aid future maintenance and make the initial installation easier. This way the installer is able to test each service and isolate quickly if a problem or leak occurs. The waste pipe is terminated within the bathroom area, so a simple connection at low level will be made.

The supply pipework emerges from the floor with the hot water supply on the left hand side. Note that the pipework is away from the side of the wall and not in the middle, as in older installations. This enables a fast and easy installation, as described later and is used on many new build installations. If the pipework is central, then moving the pipework to the side is an alternative you might consider, to ease installation.

Fix either chrome or brass isolating valves of the quarter turn ball type. Non return valves may sometimes also be fitted to prevent contamination to potable wholesome water supplies in direct water systems. The types used are compression fittings, and are installed using methods described in Joining methods (see pages 31–36). At this point and ensuring that the fittings are tight and that the valves are turned off and are facing in a position where they can be operated with ease, the water can be turned back on and tested up to the valves.

8 Before fitting the bath, make up the supply pipework to the taps using a 22 mm street elbow, a straight 22 mm tap connector, a 22 mm elbow and a short length of 22 mm pipe (8a). Alternatively, a machine bend is made in the copper tube and connected to a straight tap connector. These are soldered up with lengths of 22 mm pipe before connecting to the bath and should be long enough to protrude past the side of the bath (8b).

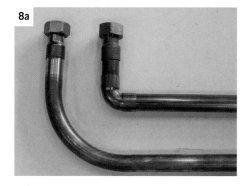

8a

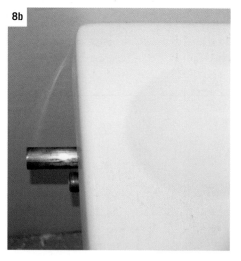

8b

9 Once the fittings have cooled down, connect the soldered pipework, complete with the tap connector to the bath. First lightly apply potable jointing compound to the seating of the tap connector before fitting the fibre washer (9a). Apply a second light coating on top of the fibre washer to stop the washer falling off when connecting it to the taps. This also helps create an excellent waterproof seal between the tap connector and the tap. Connect the first soldered pipework to the tap nearest the bath panel, and ensure the pipe protrudes past the bath edge. Tightening the tap connector is a relatively simple task now, with the use of an adjustable spanner (9b) or tap spanner. Once the pipework is secure repeat the process with the other tap, so that both pipes face towards the front of the bath where the bath panel is situated.

10 Once all these components are in place measure the distance from the back of the bath (tap end) to the leg positions and mark these clearly on the floor. Fix these securely to the floor taking care not to fix into pipework or electrical wiring which may be hidden under the floor.

Before placing the plastic bath on to the boards it is recommended you remove the centre support foot (leave the bracket attached) and replace it after the other four feet have been secured.

11 If you have levelled all the legs correctly the bath should be very close to being level when you sight it, assuming the floor is also level. Make any minor adjustment to level as required (11a), check that the bath is straight with the aid of a level, then mark the wall with a pencil along the top of the bath. Also mark the bracket positions (plastic baths only). Remove the bath, drill and plug the wall (or cavity fittings) for fixing the brackets.

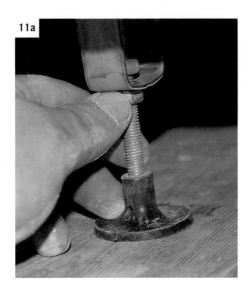

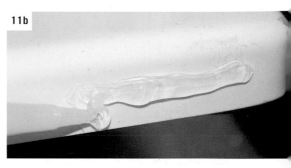

Before replacing the bath, apply a silicone sealant around the edges of the bath (11b) that will be in contact with the walls, and then refit ensuring the bath lines up with the pencil marks on the wall.

12 Once you are satisfied that the bath is level, secure it to the wall using the brackets and either cavity or screw and wall plug fittings. Now the bath is fully secured to the wall screw the supporting feet onto the boards replace and adjust the middle leg to aid support in the centre.

13 Measure from the isolation valves to the bath pipework (13a), and cut and prepare a suitable length of 22 mm pipe, taking into account the size of the fitting used (for instance a shorter length of pipework is required if a compression fitting is used). Fit the cut piece of 22 mm pipe into the isolation valve and mark and trim the bath pipework (13b).

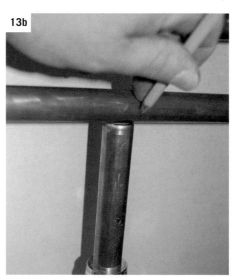

14 If using a soldered fitting then prepare pipework and fittings as described in Joining methods (see pages 31–36), and assemble (14a), fully tightening the pipe into the isolation valve. Solder the 22 mm bend; heat protection must be used (14b) and extra care should be taken to protect the bath or surrounding areas from the flames and heat of the torch with a heat mat. Repeat the process for the other side, ensuring that all flux is removed from soldered joints and that all compression fittings are fully tightened.

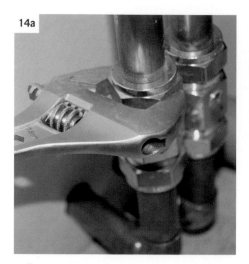

15 Connect the bath trap to the waste (15a) and measure the plastic pipework that is required. Cut the required length of pipes and prepare, using wire wool and a propriety cleaner. Disconnect the bath trap and connect the prepared pipe, and bend. Solvent-weld the piping to the fitting (15b). An alternative is either a compression or push fit fitting, but push fit fittings do not connect to all plastic pipework. Reconnect the bath trap, then connect the overflow corrugated pipe (15c).

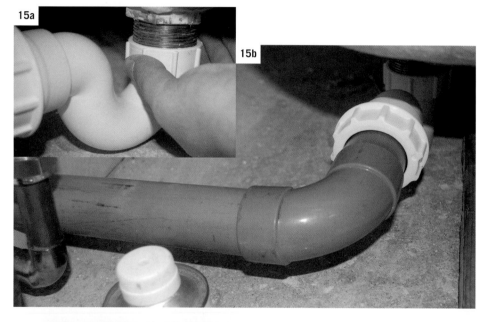

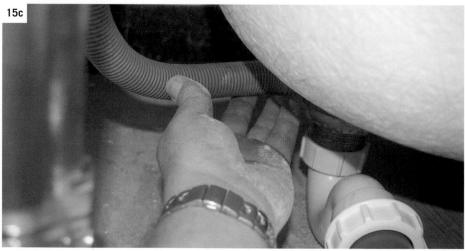

16 Now the bath has been installed, earth continuity connections must be ensured and if you are fitting a steel bath, then the bath must also be earthed along with all supply (including metal wastes) pipework (16a). Once the bath has been installed (16b) it must be fully tested. Always test up to the overflow level to ensure that it is watertight, and that the waste can actually withstand the high pressure of water once the bath is full, and that the stability of the bath is sufficient to handle that amount of water.

16a

16b

Once the testing is satisfactory the bath panel can be fitted. If it is a plastic panel this is generally fixed by clips supplied with the panel that clip onto the front legs and under the front and side lips of the bath and then screw fixed to a baton placed along the floor. If a solid board bath panel is used then a frame has to be made and the bath panel has to be screwed on with screw cups and cover caps to enable the panel to be removed easily for maintenance.

WC – TOILET CISTERN AND PAN

These are basic installation procedures for a typical close coupled and low level WC. Copper is used in this installation but all supply pipework can be installed with the use of plastic pipework and fittings.

Please note: All tasks should be installed in line with current water regulations.

For a typical close coupled WC installation, the following materials are required: (a large number of these materials should be included with the WC and cistern, and they should always be checked when purchasing a new WC prior to installation): ball valve, float, siphon, handle assembly (either push or standard side handle), (optional overflow or blanking plug), pan connection bolts and plate, rubber 'doughnut' sealing washer, 15 mm straight tap connector, 15 mm copper tube, isolation valve, screws and plastic wall plugs, lead-free solder, flux, pan connector, jointing compound, silicone lubricant, silicone sealant, earth straps and earthing wire.

For a typical low level WC installation, the following materials are required: (again, a large number of these materials should be included with the WC and cistern, and they should always be checked when purchasing a new WC prior to installation): ball valve, float, siphon, handle assembly (either push or standard side handle), flush pipe, internal or external flush cone, supporting brackets (china only), (optional overflow or blanking plug), 15 mm straight tap connector, 15 mm copper tube, isolation valve, screws and plastic wall plugs, lead-free solder, flux, pan connector, jointing compound, silicone lubricant, silicone sealant, earth straps and earthing wire.

11 Refit the pipe into the isolation valves first but don't fully tighten the nut, just hand tighten ensuring at least two threads of the isolation valve are covered. Fit the tap connector onto the ball valve ensuring the fibre washer is still in place, then just using your hand to tighten and ensuring that the plastic connector is not cross threading, tighten until the connector is on as far as it will go by hand. Using an adjustable wrench or basin wrench fully tighten the tap connector then fully tighten the isolation valve (11a). Fit the push button rods in place (11b) in the cistern lid. These may require trimming to fit.

11a

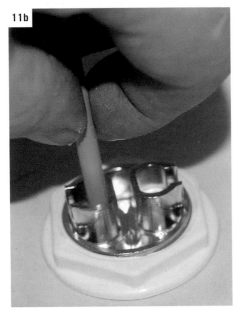
11b

12 Now the WC has been installed earth continuity connections must be made (see Earthing, pages 20–21).

Once this is done you need to test for water tightness, that the ball valve operates smoothly and that the water level is at the correct height in the cistern. Also check that the flush of the toilet works both as half and full flush and that the contents of the pan are removed sufficiently.

LOW LEVEL WC INSTALLATION

1 For low level toilet installation the fitting of the siphon to the cistern is the same except that there is no metal plate or bolts to join the cistern to the WC pan and the particular type shown has a standard handle assembly. The ball valve is of the side entry type (1a) and a side entry overflow (1b) is required. These are installed using the methods described previously. The overflow is then run away to terminate outside the building.

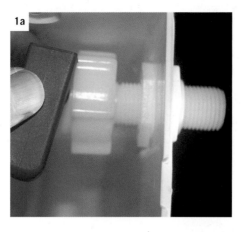

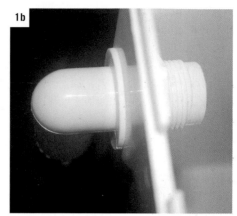

2 The handle is of a lever type and is installed by first connecting a plastic sleeve to the cistern and then inserting the handle through the sleeve, which is secured by a split pin. Place the connecting lever over the hook of the siphon (2a) and through the square shank of the handle (2b). Move the handle to ensure the siphon is working correctly. If the siphon works correctly and the handle is in the right position, then screw through the plastic connecting lever into the square shank of the handle with the screw provided (2c).

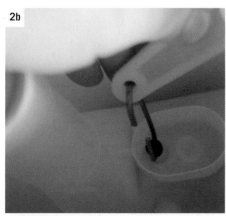

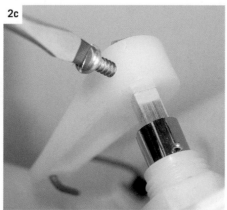

3 The low level pan is installed using the same method as the close coupled pan, with the only difference being that the distance from the wall must be further than for a close coupled pan to allow for the extra distance of the flush pipe. The height of the cistern should be approximately 850 – 900 mm and fixed either with screws through the back of the cistern or a combination of screws and brackets. With a low level WC there is a flush pipe that runs between the cistern and the pan which is secured by a nut and plastic olive onto the siphon from the cistern. Once the cistern and the pan are installed (3a), you need to determine what length of flush pipe is required to enter the WC pan (3b). Trim to suit.

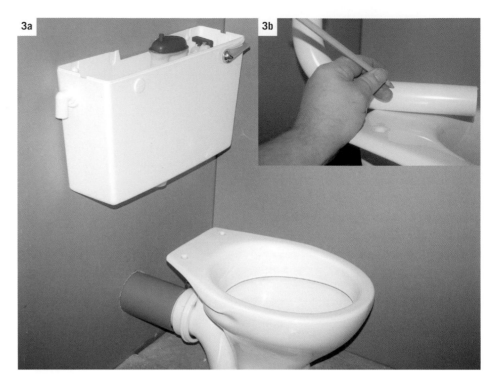

4 Connect the flush pipe to the cistern with the plastic nut and olive and to the pan by an internal cone (4a and 4b). Internal cones are normally used for most low level pans but on some an external cone is sometimes used. Now the WC has been installed, earth continuity connections must be connected. Once this is done you need to test for water tightness, that the ball valve operates smoothly and that the water level is at the correct height in the cistern. Also check that the flush of the toilet works both as half and full flush and that the contents of the pan are removed sufficiently.

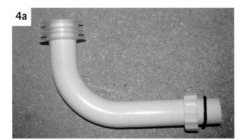

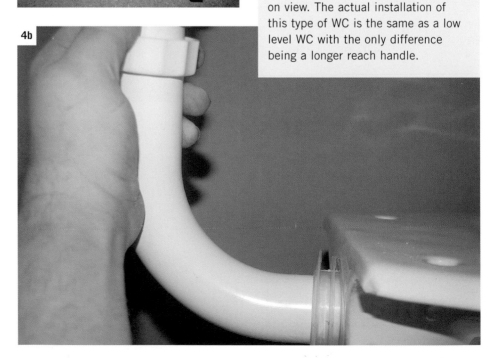

A low level cistern is a cheap way of supplying and fitting a toilet. Some toilet pans are manufactured with a flat back and are called 'back to wall pans', along with a hideaway cistern which is completely encased in boxing with only a long reach handle on view. The actual installation of this type of WC is the same as a low level WC with the only difference being a longer reach handle.

INSTALLING A COLD WATER STORAGE CISTERN (CWSC)

All cold water storage cistern installations must comply with bylaw 30 of the water regulations. Bylaw 30 is the regulation which covers the installation and storage of potable/wholesome water. Copper is used in this cold water storage cistern installation, but all pipework can be made of plastic.

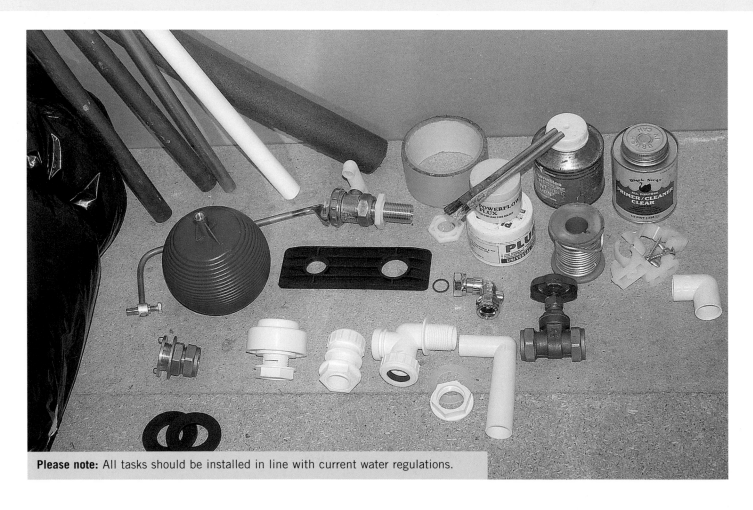

Please note: All tasks should be installed in line with current water regulations.

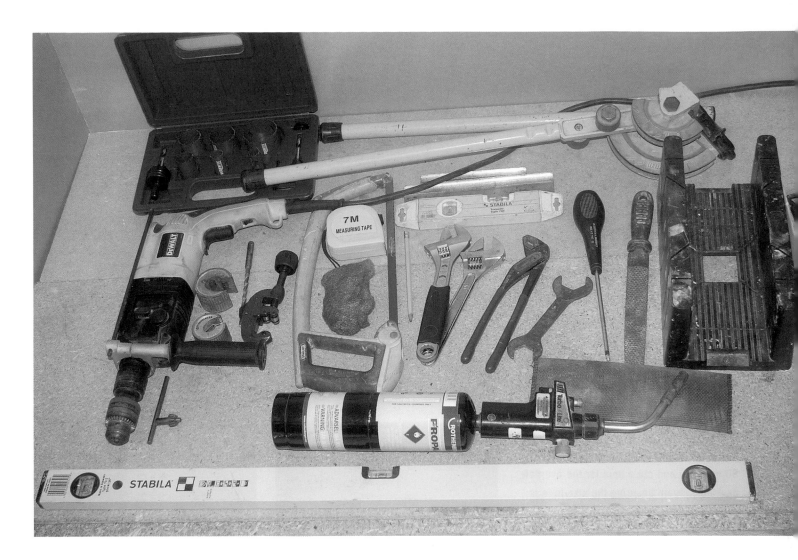

Opposite: For the installation of a typical plastic cold water storage cistern, the following materials are required: 15 mm ball valve (BS 1212 Part 2), tap connector and fibre washer, 15 mm isolation valve (or combined tap connector and isolation valve) 15 and 22 mm copper tube, 15 and 22 mm clips, 22 mm plastic overflow pipe, 22 mm plastic overflow bend, 22 mm gate valve, screws and plastic wall plugs, lead-free solder, flux, potable jointing compound, 15 and 22 mm insulation, bylaw 30 kit comprising of overflow connection, lid vent, rubber grommet for vent pipe and insulation jacket with ribbon, 22 mm tank connectors, _ in poly and rubber washers, solvent cleaner and glue, insulation sealing tape.

Above: For a typical cold water storage cistern installation, the following tools are required: two adjustable spanners, 15 and 22 mm pipe slice (or tube cutter), pipe bender (15 and 22 mm guide), long cross head screwdriver, gas torch, heat mat, wire wool, pump pliers, flat spanner, half round file, tape measure and pencil, hacksaw, drill, level, hole saw, mitre box (optional).

■ The basic preparation for this job consists of knowing the location of the cold water service valve for the rising main and its working condition. It is established that the cold water supply has been completely shut down, and within the installation of the cold water storage cistern, a new independent isolation valve of the quarter turn type is installed.

■ A selection of plastic and material type dust covers should be used with the plastic completely covering the floor area along with the material sheet on top.

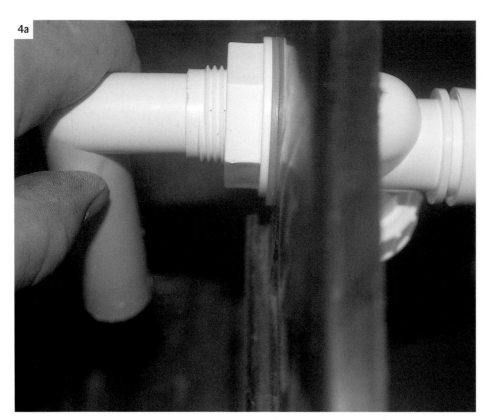

4a

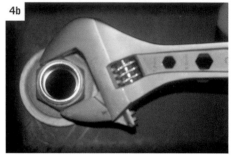

4b

4 The overflow connection can now be made. Assemble the bylaw 30 kit ensuring the insect screen is in place. Insert the overflow assembly into the second hole and place the sealing washer onto the thread. Tighten the internal plastic nut onto the thread, and insert the dip pipe into threaded end (4a). Then place the compression connector through the side wall of the cistern, using the washers provided only (4b). No jointing compound is to be used.

5 Once all the connections have been made the cistern can then be placed in the loft area. The base has to be fully supported over its whole length to ensure a solid and safe base. This base is best constructed from marine plywood with a minimum thickness of 25 mm. If a shower is to be installed in a bathroom below, then the tank will have to be raised to maintain a minimum working pressure head of 1 m. This is the distance between the base of the cistern and the shower head. Once the cistern is in place, connect the gate valve to the cold water down service pipework. These can be either wheel head or lever type, and are an essential fitting in isolating individual services.

5

6a

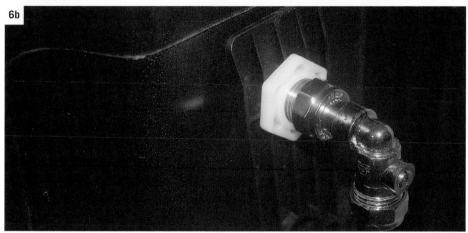

6b

6 Measure the distance required between the cold water tank connector and the gate valve (6a). Because the cistern is not in a fixed position at this stage, the initial measurements don't have to be very precise but care must be taken to maintain the stability of the cistern on the board after all connections have been made. Connect the cold water down service pipework using the cut length of 22 mm copper tube, from the compression fittings on the tank connector to the gate valve connection. Fit an isolation valve onto the rising main supply. Fit the 15 mm bent tap connector onto the ball valve; an alternative to this is a combined bent tap connector and isolation valve (6b).

7 Cut a length of copper tube to either fit between the tap connector and isolation valve or between the combined tap and isolation fitting and the rising main. Once you have fitted the tap connector and isolation valve onto the pipework connect it to the ball valve. Do this by lightly applying a potable jointing compound to the seating of the tap connector, before fitting the fibre washer and then applying a second light coating on top of the fibre washer. This stops the washer falling off when trying to connect to the valve and helps create an excellent waterproof seal between the tap connector and the valve.

7

8 If making a connection with a tap connector and isolation valve, then refit the pipe into the isolation valves first but don't fully tighten the nut, just hand tighten onto at least two threads at this stage. Then fit the tap connector onto the ball valve ensuring the fibre washer is still in place, and then just using your hand to tighten and ensuring that the connector is not cross threading, tighten until the connector is on as far as it will go by hand. Using an adjustable wrench, fully tighten the tap connector then the isolation valve.

9 The overflow pipework is the final connection to the cistern. If you are installing an overflow from new, then the following points must be adhered to: the overflow pipe should always fall away from the cistern and must not be connected to any other overflow; it must not terminate into any waste system or gutter and it should always be terminated in a place which can be easily seen, as well as well supported and insulated. A measurement is taken between the bend on the overflow pipe and the compression fitting of the overflow assembly (9a). The plastic pipework is then cut, prepared and glued onto the bend first before being fitted into the overflow assembly (9b).

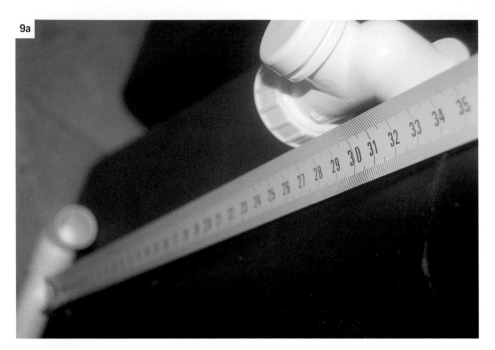

9a

9b

10 At this stage ensure the gate valves are fully closed and then test the cistern. Fill the cistern to its recommended water level, checking all connections as it fills. Once satisfied that the cistern is watertight and the ball valve is working correctly and has reached the recommended water level, the gate valves may be opened and tested, assuming the pipework is complete or valued off at another point. On the close fitting lid there will either already be holes which can be easily popped out or you may have to saw holes using the process described earlier. Connect another part of the bylaws kit (a vent with an integral insect screen) to a hole on one corner of the lid.

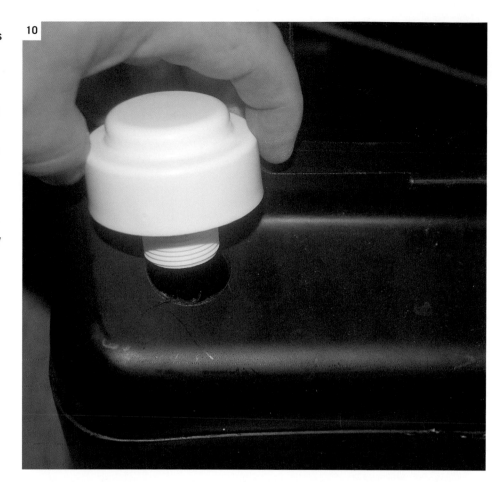

11 If the cistern is to be used with a hot water cylinder, then another hole has to be made for the vent pipe, which is used in conjunction with a plastic connector or rubber grommet, to enable a tight seal around the vent pipe. The combined hot water feed and cold water supply must be a minimum of 227 litres. This particular CWSC, however, only supplies cold water and does not require a vent pipe connection.

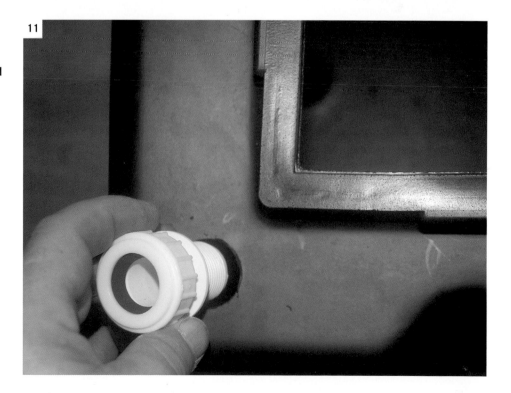

12a

12 The cold water storage cistern is now complete and once you have ensured that the close fitted lid is correctly fitted to the cistern, then all sides of the cistern, lid and pipework must be fully insulated. Use the insulation material which should be part of the bylaw 30 package and wrap it around and over the lid of the cistern and secure with the ribbon provided. The pipework should be insulated using a foam insulation along with insulation tape for the sealing of joints (12a). The base of the cistern must not be insulated. The use of a mitre block and fine-toothed hacksaw is an invaluable aid when covering pipework with insulation. The mitre allows you to make tight, close fitting angles around bends and other fittings, and used in conjunction with the sealing tape, helps maintain a well insulated system (12b). Cut out a section of insulation to enable the quarter turn isolating valve to be used (12c).

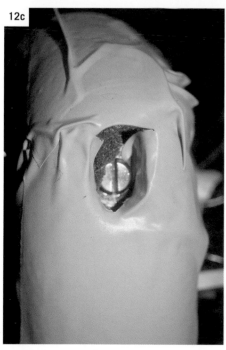

INSTALLING A KITCHEN SINK

(with washing machine and dishwasher connections)

When installing any metal kitchen sink whether it is an insert sink or a sit-on style sink, you must always take great care in the preparation and the handling of their sharp edges. These sinks can cause serious injuries by cutting into your hands during installation, so always wear strong gloves during the installation.

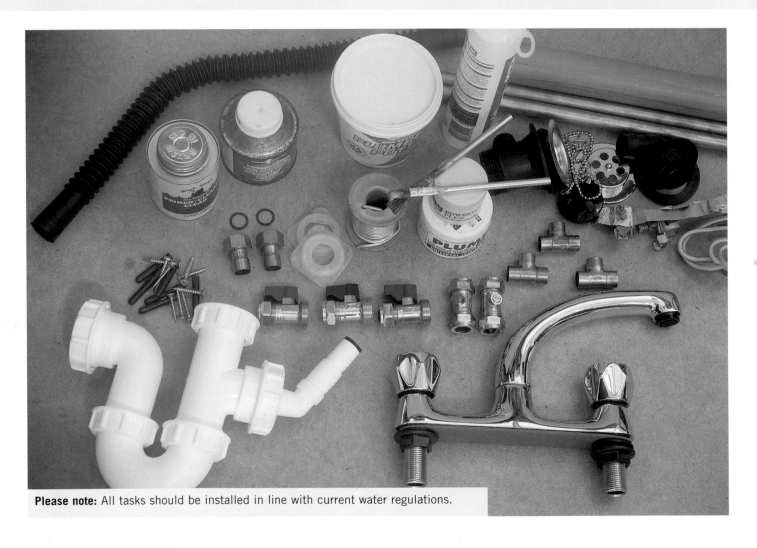

Please note: All tasks should be installed in line with current water regulations.

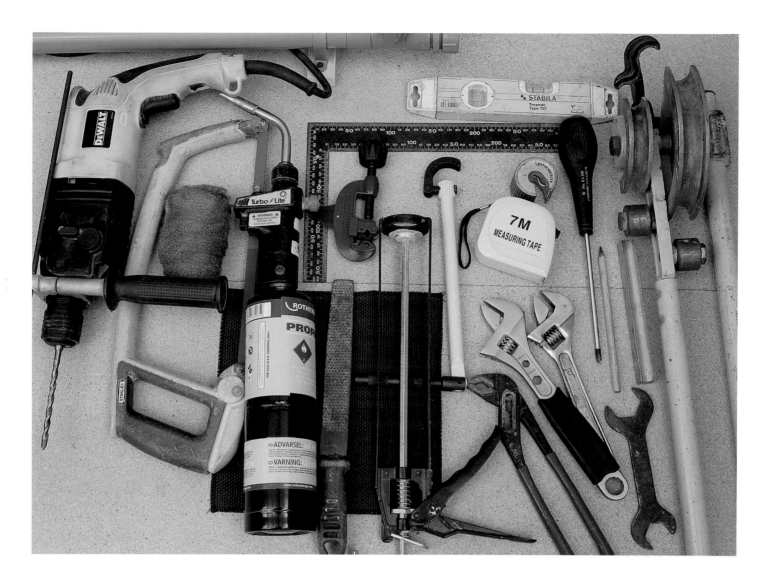

Oopposite: For a typical insert or sit-on sink kitchen sink installation, the following materials are required: combined waste and overflow assembly, plug and chain, pillar taps (alternatives are deck mixer or mono bloc), two tap connectors and fibre washers, top hats, 15 mm copper tube, 15 mm clips, 15 mm bends, 15 mm tees, isolation valves, silicone sealant, screws, wall plugs, lead-free solder, flux, 40 mm 'P' trap with combined washing machine connection, 40 mm plastic waste pipe, solvent cleaner and glue, potable wholesome jointing compound, three washing machine valves, washing machine stand pipe and trap, earth strap connections and earth wire.

Above: For a typical insert or sit-on sink kitchen sink installation, the following tools are required: two adjustable spanners, 15 mm pipe slice (or tube cutter), pipe bender (and 15 mm guide), long cross head screwdriver, gas torch, heat mat, wire wool, pump pliers, basin spanner, flat spanner, rasp or file, tape measure and pencil, hacksaw, drill, electric drill, 7 mm drill, and level.

■ The basic preparation for this job, or any project in this book, consists of knowing the location of the hot and cold service valves and their working condition, and whether it is possible to isolate either the kitchen sink or the whole of the hot and cold water supply.

■ For all the projects in this book it is established that the hot and cold water supply has been completely shut down, and within the installations of the kitchen sink new independent isolation valves of the quarter turn type are installed.

1 Remove the packaging and the plastic covering around the sink, waste and overflow areas. Apply silicon sealant or plumber's putty around the inlet waste (1a) hole on the inside of the bowl. Then push the waste firmly against the sealant and through the sink. Turn the sink over and apply more silicon sealant on the underside of the sink where the waste outlet comes through. Place the washer onto the waste along with the overflow connector (1b) and firmly secure using a screw through the chrome plug into the plastic waste and overflow connector (1c). Remove any residue of sealant with a damp cloth before proceeding with the overflow connection.

A selection of plastic and material type dust covers should be used with the plastic completely covering the floor area along with the material sheet on top.

2 Apply a small amount of silicone or plumber's putty around the overflow on the internal side of the sink, and in the same way as the waste installation, push the overflow connection through the sink and on to the sealant (2a). Apply another small amount of sealant on to the rear of the overflow connection and seal using a poly washer and threaded overflow connector (2b). Connect the corrugated overflow pipe between the waste and overflow assemblies (2c).

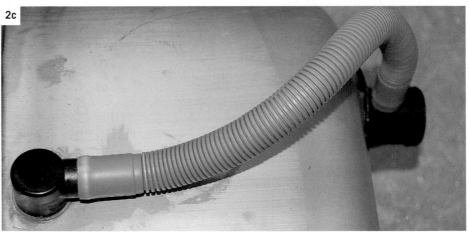

3 While the sink is still in this position push the tap through one of the tap holes ensuring that the hot tap is placed on the left hand side when the sink is facing you. A small bead of silicon or putty around the top hole of the sink before the tap is actually pushed through will help seal the tap. Where the tap protrudes through the hole underneath the sink place a top hat on to the threaded tail of the tap, then using the back nut, tighten the tap until it is in a fixed firm position (3a). Repeat the same sequence with the other tap. The sink is now connected with taps and waste assembly (3b). If a mono bloc tap is used, then the use of top hats is not necessary.

3a

3b

4 When the taps, the overflow and waste are connected remove the rest of the plastic covering. A typical pipework installation inside a kitchen carcass is shown here. As previously stated there are no independent valves for the sink, and although the water regulations don't ask for valves to be fitted to sinks it is good practice to fit isolating valves to aid future maintenance and make the initial installation easier, with the installer being able to test each service and isolate quickly if a problem or leak occurs.

5 The waste pipe is also terminated within the kitchen area, so a simple connection can be made within the carcass and an additional stand pipe and trap for a washing machine can be achieved outside the carcass area. Fix either chrome or brass isolating valves of the quarter turn ball type on the hot and cold water supply pipework inside the base unit carcass (5a). The types shown are compression fittings and are installed using methods described in Jointing methods (see pages 31–36). Once you have ensured that the fittings are tight and that the valves are turned off and are facing to the front of the unit for easy access (5b), the water can be turned back on and tested to the valves.

6 The sink top can now be connected to the supply pipes prior to actually fitting onto the unit or worktop. Connect the tap connectors onto the taps, then loosely sit the sink top onto the base unit and measure from the tap connector to the isolation valves. This measurement is only an approximation of the length of pipe required, because of the added tee fittings and the alignment of the inlet services. Cut and prepare the required lengths of 15 mm copper tube. Solder the tap connectors onto the pipe as described in Joining methods (see pages 31–36). Once the fittings have cooled down connect the tap connectors and pipe to the sink taps.

7 Lightly apply potable jointing compound to the seating of the tap connector, before fitting the fibre washer (7a). Apply a second light coating on top of the fibre washer to stop the washer falling off when trying to connect it to the taps and to help create a waterproof seal between the tap connector and the tap. Fully tighten the tap connector with an adjustable wrench (7b) or spanner. The sink is now complete with new supply pipework connected to the taps and can now be installed onto the base unit.

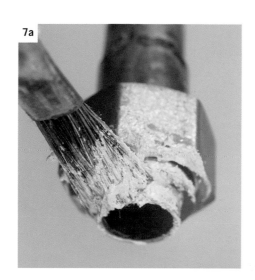

8 If installing an inserted sink into a work top, always seal the edge with a propriety sealant and use the recommended foam sealant and clips supplied with the sink. Clip the sink unit down onto the base unit with the clips provided, firmly tightening the sink down in a sequential manner.

Using the 15 mm pipe slice cut in to the cold feed pipe (8a) connected to the tap. Prepare and fit two 15 mm tees in the pipe and face one to the right, and using the pipe bender, put a slight angle on the other tail and send it to the left (8b). Measure between the cold feed pipe and the isolation valve and bend a piece of 15 mm copper to suit, as described in Bending copper tube (see pages 59–65).

9 Fit the prepared pipe into both the tee and the isolation valve (9a). Ensuring that there is adequate heat protection in the form of a heat mat, solder the tees and pipework together (9b) as described in Joining methods (see pages 31–36). When the fittings have cooled down tighten the isolation valve. Repeat the process for the hot water supply pipe, but only fit one tee and face it to the left alongside the cold branch.

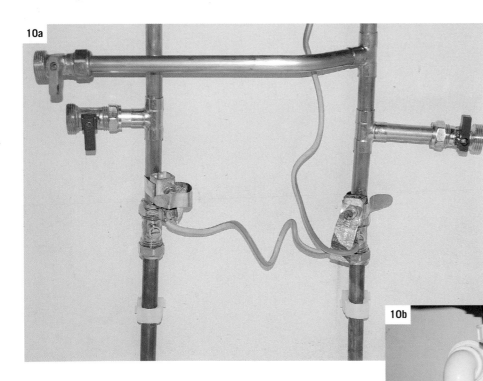

10 Fit the three washing machine valves, making sure they have the correct colour handle for the hot (red) and cold (blue) and that they are fully tightened and in the off position. Dishwashers generally have only a cold feed but washing machines can have either hot and cold or cold fill only. Earthing connections must be maintained especially from the kitchen sink and across the hot and cold services (10a), as described in Earthing (see pages 20–21). Connect the 'P' trap with a combined washing machine spigot to the waste outlet of the sink (10b).

11 The waste pipe in this installation terminates through the rear of the base unit, so a measurement is taken from the extruding waste pipe into the full depth of the trap, marked and cut in situ. If the waste enters in an offset position, then a combination of bends is used; these can be either glued, push fit or compression. After cleaning and de-burring the waste pipe as described in Joining methods (see pages 31–36) connect the trap onto the plastic waste pipe and then onto the sink waste.

12 The spigot connection which is part of the sink trap could be used for a dishwasher or washing machine. The spigot end is either cut or has a removable cap (12a) to the required diameter of the corrugated waste pipe of the machine being used, and is secured by means of a jubilee clip. The kitchen sink is now complete (12b).

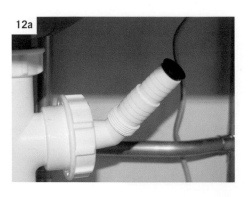

12a

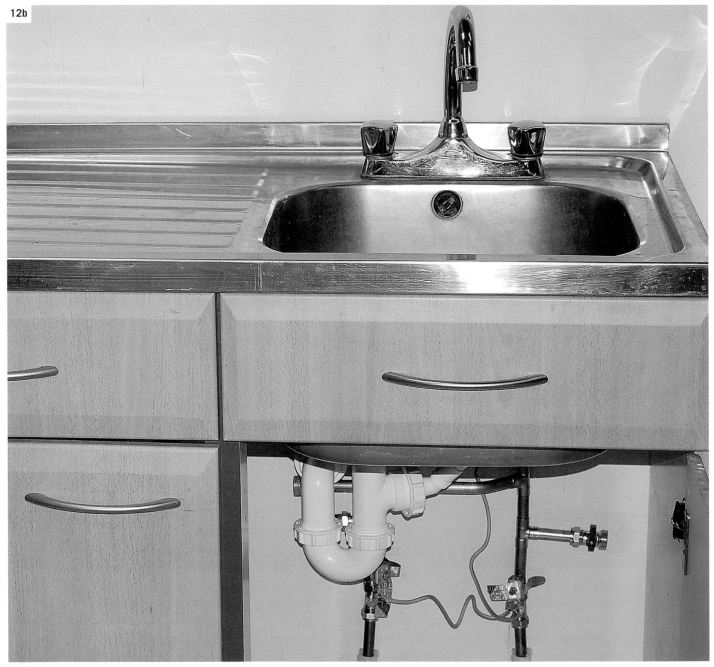

12b

13 Outside the base unit, a tee has been cut into the 40 mm waste pipe to allow for an additional washing machine trap to be fitted (13a). Notice that the open branch of the tee allows for the flow to run with the waste pipe. Clip the stand pipe and trap onto the wall in a place where it will not cause an obstruction to the washing machine being set back to the wall (13b). Measure, prepare and insert a small length of pipe out of the tee and into a bend (13c). Measure from the bend into the trap, cut and prepare pipe. Glue the bend onto the pipes (marking which pipe goes where), then glue the assembly into the tee and finally insert in the compression fitting of the trap. Test both hot and cold water supplies individually and let the water run until it runs clear. Test the new trap and waste system with the plug in and out. Also test the new washing machine trap outside the base unit.

13a

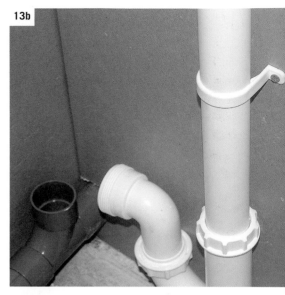
13b

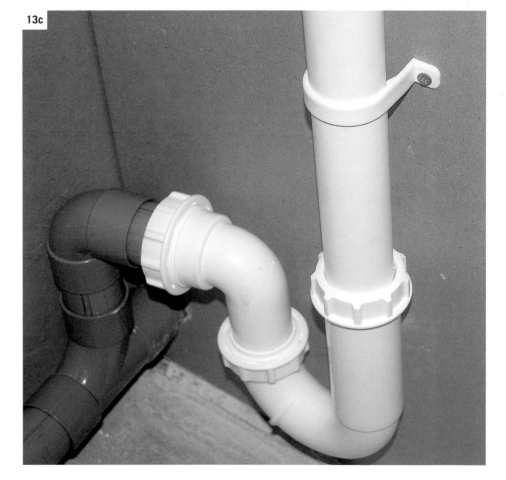
13c

INSTALLING AN OUTSIDE TAP

This is a typical installation of an outside tap comprising a simple screw down bib cock garden tap. Even for a job as small as this it is best to allow yourself an hour's preparation time before attempting it.

For a typical outside tap installation, the following materials are required: back plate elbow, screw down bib cock with internal double check valve, 15 mm copper tube, 22 mm copper tube, 15 mm clips, 15 mm bends, isolation valves, sealant, screws and plastic wall plugs, 15 mm equal tee, lead-free solder, flux, PTFE potable/wholesome jointing compound.

For a typical outside tap installation, the following tools are required: two adjustable spanners, 15 mm pipe slice, tube cutter, long cross head screwdriver, gas torch, heat mat, wire wool, pump pliers, flat spanner, tape measure and pencil, electric drill, 7 mm drill, 15 mm and 22 mm masonry drills, level.

■ The basic preparation for this job consists of knowing the location of the cold water service valves and their working condition, and whether it is possible to isolate either the cold water supply you are cutting into (kitchen cold water supply) or the complete cold water supply of the property.

■ A selection of plastic and material type dust covers should be used inside the property with the plastic completely covering the floor area along with the material sheet on top.

1 Start by determining where you are going to connect the tap and that it will not cause a problem or obstruction, and will not be too close to any electrical points. If possible it is always preferable to site an outside tap over a gulley or drain. Once you have established where you are going to site the tap and where it is going to be fed from, the next stage is to mark the outside of the wall before drilling (1). Ensure there are no hidden pipework or electrical wires in the way first by using a pipe and electric cable detector.

2 Measure the thickness of the wall to be drilled and mark this depth on the 15 mm drill with a small piece of electrical or masking tape. Then using the 15 mm masonry drill on a hammer selection and drilling from the outside in drill until approximately 50 mm from the tape, at this point turn the hammer selection off. Now drill the final 50 mm on drill only, this should hopefully restrict any damage to a minimum, and produce a smooth hole. Once the 15 mm hole has been achieved, and using the 22 mm masonry bit with the drill on the basic drilling selection, carefully drill from the inside back through the hole to approx half the depth of the hole, and repeat the procedure from the outside drilling back.

3 The hole is now large enough to accept the 22 mm sleeving which should be cut with a hacksaw (to enable the bend to fit into the end of the sleeve) from the measurement taken in Step 2. Insert the length of 22 mm pipe work into the hole. If installing an outside tap that does not have a double check valve incorporated in its manufacture, then you must fit a double check valve in the supply pipework to the tap to comply with current water regulations.

4 Wrap the PTFE thread sealing tape around the bib cock (4a) and screw onto the back plate elbow, ensuring the tap tightens at the correct angle to the back plate (4b and 4c). Additional PTFE may be needed to create the correct angle. Place the back plate elbow onto the wall, and measure the length of 15 mm copper tube from the back plate elbow through the wall and allowing enough tube to make a connection to the supply. Cut and prepare the lengths of 15 mm copper tube along with an end feed elbow. Solder the elbow onto the pipe as described in Joining methods (see pages 31–36).

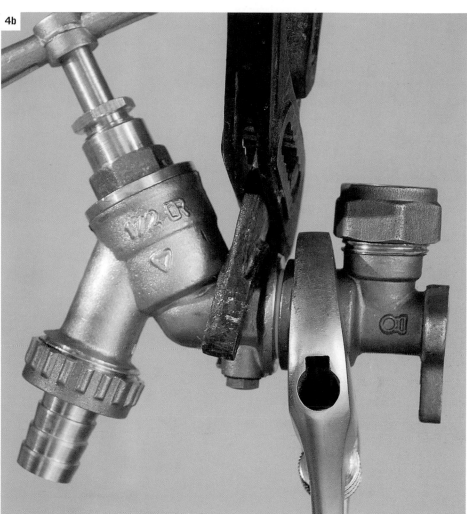

4b

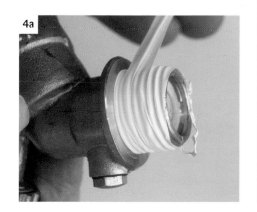

4a

4c

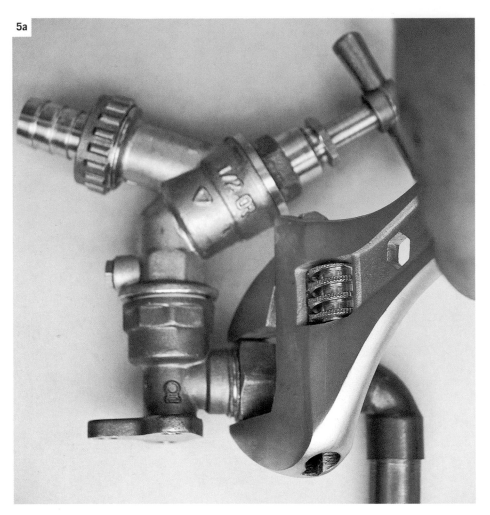

5a

5 Once the fitting and pipe have cooled down connect the pipe to the back plate elbow (5a). This is achieved by a simple compression fitting as described in Joining methods (see pages 31–36). Push the tap and pipe assembly through the sleeved hole (5b) and mark the fixings for the back plate elbow. Drill and plug the wall.

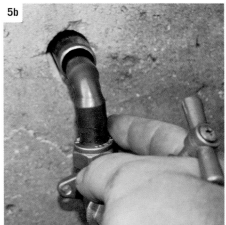

5b

6 Replace the tap and pipe assembly back through the sleeved hole and fix the tap to the wall by screwing into the plugs through the back plate elbow. Inside the cupboard area where the pipe has terminated, measure the distance between the supply and the new pipework (6a). Cut and prepare a length of 15 mm copper tube to this measurement, complete with an isolation valve. Place the prepared tube in place and mark both the supply pipe and new pipe for cutting (6b). Ensure all earthing connections are maintained. Once you have ensured the water supply has been turned off and using a pipe slice cut both these pieces of pipework and prepare for soldering.

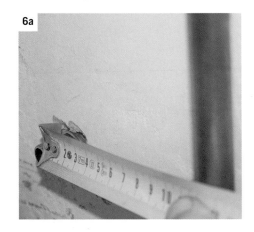

6a

6b

7 Assemble the pipework, tee and elbow together inside the base unit, and ensuring that there is adequate heat protection, solder the tee, bend and pipework together as described in Joining methods (see pages 31–36) (7a). Once the fittings have cooled down the pipework and tap can be tested to ensure a water tight installation. Once this is achieved, the sleeving can be sealed with a proprietary sealant around the supply pipe (7b and 7c) to the tap, both inside and out.

7a

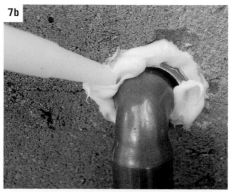

7b

7c

PROJECT 7

INSTALLING A HOT WATER CYLINDER

This is a typical installation of a basic direct hot water cylinder consisting of a typical immersion heater assembly. The hot water should only be heated to between 60–65° C (140–149° F). A hot water cylinder should only be installed by a someone who is confident in their ability to carry out this task

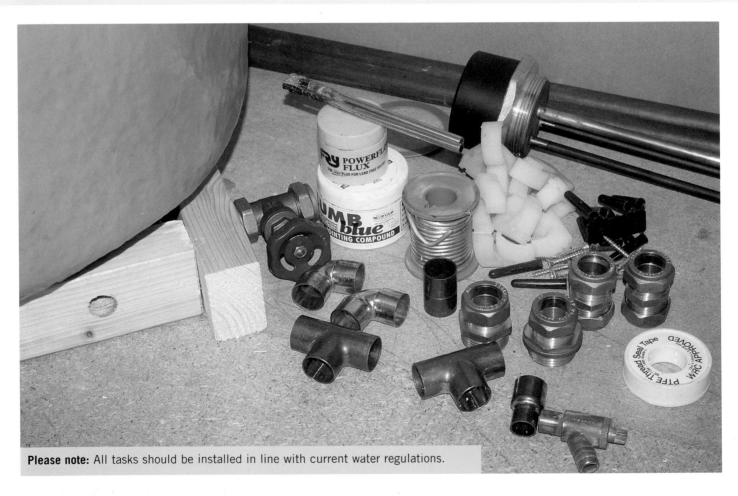

Please note: All tasks should be installed in line with current water regulations.

For a typical hot water cylinder installation, the following materials are required: 22 mm copper tube, 22 mm clips, two 22 mm bends, two 22 mm tees, two equal tees, two

22 mm compression couplings, two 1 inch male iron to 22 mm copper straight connectors, drain off cock (DOC), 22–15 mm internal reducer, screws and plastic wall

plugs, lead-free solder, flux, PTFE, potable jointing compound, immersion heater, washer and thermostat, gate valve, wooden base for cylinder.

For a typical hot water cylinder installation, the following tools are required: two adjustable spanners, 22 mm pipe slice (or tube cutter), pipe bender (and 22 mm guide), long cross head screwdriver, gas torch, heat mat, wire wool, pump pliers, flat spanner, tape measure and pencil, drill, 7 mm masonry drill, hole saw, pad saw, level, immersion spanner.

■ For this project it is established that the cold feed to the cylinder is sited at high level, and the hot water supply from the cylinder is located at low level.

■ A selection of plastic and material type dust covers should be used with the plastic completely covering the floor area along with the material sheet on top.

■ This task is an installation of a direct cylinder which is heated via an emersion heater. If the direct cylinder has additional tapings for either a gravity hot water system or secondary return, then these should be capped off using a suitable brass plug. Before commencing installation of the direct cylinder a small stand should be made enabling an air flow under the cylinder.

■ With this type of direct cylinder installation there are only two connections to be made along with the connection for the immersion heater. For this project the CWSC is sited directly above the cylinder cupboard for ease of installation, but in many other properties additional lengths of pipework may have to be used.

1 Place the cylinder onto the wooden base in the cylinder cupboard, ensuring there is enough clearance for all the pipework connections to be made, and that the cylinder is both stable and level (1a). With the cylinder and base still in place and using a level, mark a line from the cold feed pipe at high level towards the floor (1b). Fit the 22 mm clips onto this line taking care to leave enough space at both high and low connections (1c).

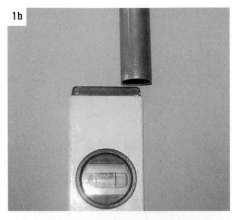

2 Repeat the same procedure for the hot water feed pipe, but this time mark a level line from the hot water supply pipe at low level towards the ceiling, ensuring a minimum distance of 450 mm from the centre mark of the hot water outlet. Fit 22 mm clips in the same way. This may entail moving the pipework further to the side, by use of fittings or machine bending.

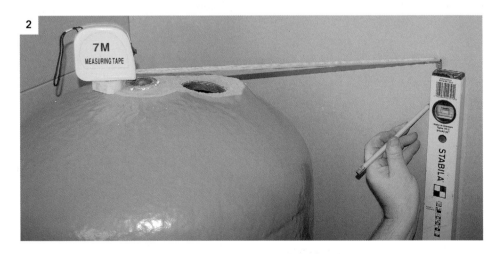

3 Once all the clips are in place, place a small length of copper tube into the hot water feed clips at high level, and mark around the pipe on the ceiling (3a). The hole in the ceiling can now be cut out by using either a pad saw or hole saw (3b). Care must be taken that no pipework or electrical wires are present before cutting this hole. You may also have to remove the clips at high level to accomplish this.

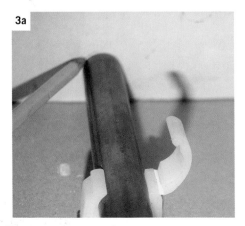

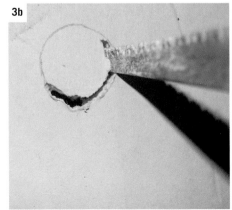

4 Measure down from the cold feed pipe to a position at high level but between clips (4a), and cut and prepare a length of 22 mm copper tube to this measurement. Connect the straight coupling compression fitting to one end of the copper tube and a gate valve to the other. These are installed using methods described in Jointing methods (see pages 31–36). Fit the assembled length of copper tube in place, by connecting the straight coupling compression fitting onto the cold feed pipe (4b and 4c). I have found that it is much safer to use compression fittings on pipework in areas that are close to ceilings, floors or ductwork. If using a gas torch on soldered fittings, the flame from the torch can be drawn into the hole around the pipe, causing damage or possibly even a fire.

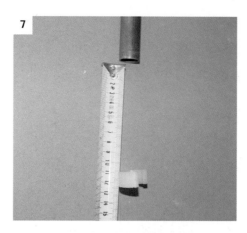

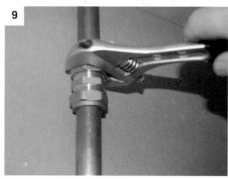

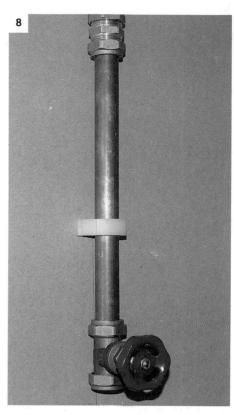

5 Measure from the gate valve to just below the mark on the wall for the cold feed inlet. Cut and prepare a length of 22 mm copper tube to this dimension and fit it into the gate valve and clips. Fully tighten gate valve. Measure the distance between the hot water feed and the cylinder height mark on the wall, and cut and prepare a length of copper tube to this dimension. Connect the straight coupling compression fitting to one end of the copper tube and fit onto the hot water feed pipe and into the clips.

6 Fit the hot water outlet connection which is situated at the top of the cylinder. This is made with a 1 inch male iron to 22 mm copper fitting along with PTFE thread sealing tape. The PTFE is wound around the thread of the fitting as described in Joining methods (see pages 31–36) (6a). The fitting is then screwed into the cylinder by hand at first, to ensure cross threading does not occur. It is then securely tightened using an adjustable wrench (6b).

6a

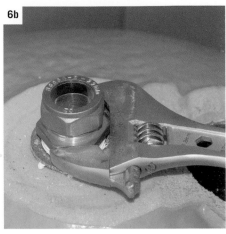

6b

7a

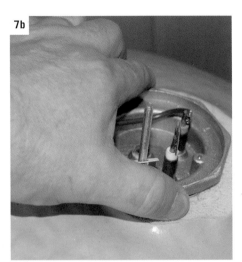

7b

7 Repeat the same procedure for the cold feed inlet connection. Wind the PTFE sealing tape around the thread of the immersion heater (7a). The immersion must be of a sufficient length to enable the cylinder to be heated economically. Ensuring the large fibre washer is firmly in place on the immersion, screw the immersion onto the cylinder by hand at first to ensure cross threading does not occur (7b). Then securely tighten using an immersion spanner (7c) and insert the immersion thermostat (7d).

7c

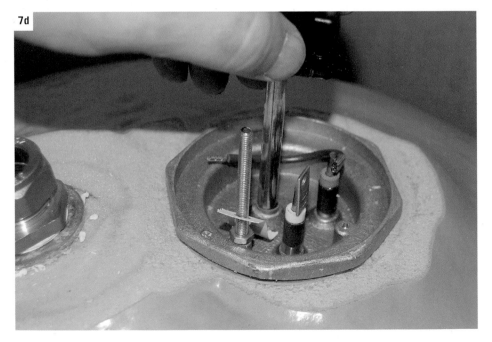

7d